KU-490-427

Primary SPACE Project Research Team

Research Co-ordinating Group

Professor Paul Black (Co-director)
Jonathan Osborne

Centre for Educational Studies
King's College London
University of London
Cornwall House Annexe
Waterloo Road
London SE1 8TZ

Tel: 071 872 3094

Dr Wynne Harlen (Co-director)
Terry Russell

Centre for Research in Primary Science and Technology
Department of Education
University of Liverpool
126 Mount Pleasant
Liverpool L3 5SR

Tel: 051 794 3270

Project Researchers

Pamela Wadsworth (from 1989)

Derek Bell (from 1989)
Ken Longden (from 1989)
Adrian Hughes (1989)
Linda McGuigan (from 1989)
Dorothy Watt (1986-89)

Associated Researchers

John Meadows
(South Bank Polytechnic)

Bert Sorsby
John Entwistle
(Edge Hill College)

LEA Advisory Teachers

Maureen Smith (1986-89)
(ILEA)

Joan Boden
Karen Hartley
Kevin Cooney (1986-88)
(Knowsley)

Joyce Knaggs (1986-88)
Heather Scott (from 1989)
Ruth Morton (from 1989)
(Lancashire)

PRIMARY SPACE PROJECT
RESEARCH REPORT

January 1990

Sound

by
DOROTHY WATT and TERRY RUSSELL

LIVERPOOL UNIVERSITY PRESS

First published 1990 by
Liverpool University Press
PO Box 147, Liverpool L69 3BX

Reprinted, with corrections, 1992

British Library Cataloguing in Publication Data
Data are available
ISBN 0 85323 456 6

Printed and bound by
Antony Rowe, Limited, Chippenham, England

CONTENTS

INTRODUCTION

This introduction is common to all SPACE topic reports and provides an overview of the Project and its programme.

The Primary SPACE Project is a classroom-based research project which aims to establish

- *the ideas which primary school children have in particular science concept areas*
- *the possibility of children modifying their ideas as the result of relevant experiences.*

The research is funded by the Nuffield Foundation and is being conducted at two centres, the Centre for Research in Primary Science and Technology, Department of Education, University of Liverpool and the Centre for Educational Studies, King's College, London. The joint directors are Professor Wynne Harlen and Professor Paul Black. The Project has one full-time researcher, based in Liverpool, and is supported by a range of other personnel (refer to Project team). Three local education authorities are involved: Inner London Education Authority, Knowsley and Lancashire.

The Project is based on the view that children develop their ideas through the experiences they have. With this in mind, the Project has two main aims: firstly, to establish (through an elicitation phase) what specific ideas children have developed and what experiences might have led children to hold these views; and secondly, to see whether, within a normal classroom environment, it is possible to encourage a change in the ideas in a direction which will help children develop a more 'scientific' understanding of the topic (the intervention phase).

Eight concept areas have been studied:

Electricity
Evaporation and condensation
Everyday changes in non-living materials
Forces and their effect on movement
Growth
Light
Living things' sensitivity to their environment
Sound.

The Project has been run collaboratively between the University research teams, local education authorities and schools, with the participating teachers playing an active role in the development of the Project work.

Over the two year life-span of the Project a close relationship has been established between the University researchers and the teachers, resulting in the development of techniques which advance both classroom practice and research. These methods provide opportunities, within the classroom, for children to express their ideas and to develop their thinking with the guidance of the teacher, and also help researchers towards a better understanding of children's thinking.

The involvement of teachers

Schools and teachers were not selected for the Project on the basis of a particular background or expertise in primary science. In the majority of cases, two teachers per school were involved, which was advantageous in providing mutual support. Where possible, the Authority provided supply cover for the teachers so that they could attend Project sessions for preparation, training and discussion during the school day. Sessions were also held in the teachers' own time, after school.

The Project team aimed to have as much contact as possible with the teachers throughout the work to facilitate the provision of both training and support. The diversity of experience and differences in teaching style which the teachers brought with them to the Project meant that achieving a uniform style of presentation in all classrooms would not have been possible, or even desirable. Teachers were encouraged to incorporate the Project work into their existing classroom organisation so that both they and the children were as much at ease with the work as with any other classroom experience.

The involvement of children

The Project involved a cross-section of classes of children throughout the primary age range. A large component of the Project work was classroom-based, and all of the children in the participating classes were involved as far as possible. Small groups of children and individuals were selected for additional activities or interviews to facilitate more detailed discussion of their thinking.

The structure of the Project

For each of the eight concept areas studied, a list of concepts was compiled to be used by researchers as the basis for the development of work in that area. These lists were drawn up from the standpoint of accepted scientific understanding and contained concepts which were considered to be a necessary part of a scientific understanding of each topic. The lists were not necessarily considered to be statements of the understanding which would be desirable in a child at age eleven, at the end of the Primary phase of schooling. The concept lists defined and outlined the area of interest for each of the studies; what ideas children were able to develop was a matter for empirical investigation.

Most of the Project work can be regarded as being organised into four phases, preceded by an extensive pilot phase. These phases are described in the following paragraphs and are as follows:

Pilot work
Phase 1: Exploration
Phase 2: Pre-Intervention Elicitation
Phase 3: Intervention
Phase 4: Post-Intervention Elicitation

The phases of the research

Each phase, particularly the Pilot work, was regarded as developmental; techniques and procedures were modified in the light of experience. The modifications involved a refinement of both the exposure materials and the techniques used to elicit ideas. This flexibility allowed the Project team to respond to unexpected situations and to incorporate useful developments into the programme.

There were three main aims of the Pilot phase. Firstly, to trial the techniques used to establish children's ideas; secondly, to establish the range of ideas held by primary school children; and thirdly, to familiarise the teachers with the classroom techniques being employed by the Project. This third aim was very important since teachers were being asked to operate in a manner which, to many of them, was very different from their usual style. By allowing teachers a 'practice run', their initial apprehensions were reduced, and the Project rationale became more familiar. In other words, teachers were being given the opportunity to incorporate Project techniques into their teaching, rather than having them imposed upon them.

In the Exploration phase children engaged with activities set up in the classroom for them to use, without any direct teaching. The activities were designed to ensure that a range of fairly common experiences (with which children might well be familiar from their everyday lives) was uniformly accessible to all children to provide a focus for their thoughts. In this way, the classroom activities were to help children articulate existing ideas rather than to provide them with novel experiences which would need to be interpreted.

Each of the topics studied raised some unique issues of technique and these distinctions led to the Exploration phase receiving differential emphasis. Topics in which the central concepts involved long-term, gradual changes, e.g. 'Growth', necessitated the incorporation of a lengthy exposure period in the study. A much shorter period of exposure, directly prior to elicitation was used with 'Light' and 'Electricity', two topics involving 'instant' changes.

During the Exploration, teachers were encouraged to collect their children's ideas using informal classroom techniques. These techniques were:

i. *Using log-books (free writing/drawing)*

Where the concept area involved long-term changes, it was suggested that children should make regular observations of the materials, with the frequency of these depending on the rate of change. The log-books could be pictorial or written, depending on the age of the children involved, and any entries could be supplemented by teacher comment if the children's thoughts needed explaining more fully. The main purposes of these log-books were to focus attention on the activities and to provide an informal record of the children's observations and ideas.

ii. *Structured writing/drawing*

Writing or drawings produced in response to a particular question were extremely informative. This was particularly so when the teacher asked children to clarify their diagrams and themselves added explanatory notes and comments where necessary, after seeking clarification from children.

Teachers were encouraged to note down any comments which emerged during dialogue, rather than ask children to write them down themselves. It was felt that this technique would remove a pressure from children which might otherwise have inhibited the expression of their thoughts.

iii. *Completing a picture*

Children were asked to add the relevant points to a picture. This technique ensured that children answered the question posed by the Project team and reduced the possible effects of competence in drawing skills on ease of expression of ideas.

iv. *Individual discussion*

The structured drawing provided valuable opportunities for teachers to talk to individual children and to build up a picture of each child's understanding.

It was suggested that teachers use an open-ended questioning style with their children. The value of listening to what children said, and of respecting their responses, was emphasised as was the importance of clarifying the meaning of words children used. This style of questioning caused some teachers to be concerned that, by accepting any response whether right or wrong, they might implicitly be reinforcing incorrect ideas. The notion of ideas being acceptable and yet provisional until tested was at the heart of the Project. Where this philosophy was a novelty, some conflict was understandable.

In the Elicitation phase, the Project team collected structured data through individual interviews and work with small groups. The individual interviews were held with a random, stratified sample of children to establish the frequencies of ideas held. The same sample of children was interviewed pre- and post-Intervention so that any shifts in ideas could be identified.

The Elicitation phase produced a wealth of different ideas from children, and led to some tentative insights into experiences which could have led to the genesis of some of these ideas. During the Intervention teachers used this information as a starting point for classroom activities, or interventions, which were intended to lead to children extending their ideas. In schools where a significant level of teacher involvement was possible, teachers were provided with a general framework to guide their structuring of classroom activities appropriate to their class. Where opportunities for exposing teachers to Project techniques were more limited, teachers were given a package of activities which had been developed by the Project team.

Both the framework and the Intervention activities were developed as a result of preliminary analysis of the Pre-Intervention Elicitation data. The Intervention strategies were:

(a) Encouraging children to test their ideas

It was felt that, if pupils were provided with the opportunity to test their ideas in a scientific way, they might find some of their ideas to be unsatisfying. This might encourage the children to develop their thinking in a way compatible with greater scientific competence.

(b) Encouraging children to develop more specific definitions for particular key words

Teachers asked children to make collections of objects which exemplified particular words, thus enabling children to define words in a relevant context, through using them.

(c) Encouraging children to generalise from one specific context to others through discussion.

Many ideas which children held appeared to be context-specific. Teachers provided children with opportunities to share ideas and experiences so that they might be enabled to broaden the range of contexts in which their ideas applied.

(d) Finding ways to make imperceptible changes perceptible

Long-term, gradual changes in objects which could not readily be perceived were problematic for many children. Teachers endeavoured to find appropriate ways of making these changes perceptible. For example, the fact that a liquid could 'disappear' visually and yet still be sensed by the sense of smell - as in the case of perfume - might make the concept of evaporation more accessible to children.

(e) Testing the 'right' idea alongside the children's own ideas

Children were given activities which involved solving a problem. To complete the activity, a scientific idea had to be applied correctly, thus challenging the child's notion. This confrontation might help children to develop a more scientific idea.

In the Post-Intervention Elicitation phase the Project team collected a complementary set of data to that from the Pre-Intervention Elicitation by re-interviewing the same sample of children. The data were analysed to identify changes in ideas across the sample as a whole and also in individual children.

These four phases of Project work form a coherent package which provides opportunities for children to explore and develop their scientific understanding as a part of classroom activity, and enables researchers to come nearer to establishing what conceptual development it is possible to encourage within the classroom and the most effective strategies for its encouragement.

The implications of the research

The SPACE Project has developed a programme which has raised many issues in addition to those of identifying and changing children's ideas in a classroom context. The question of teacher and pupil involvement in such work has become an important part of the Project, and the acknowledgement of the complex interactions inherent in the classroom has led to findings which report changes in teacher and pupil attitudes as well as in ideas. Consequently, the central core of activity, with its pre- and post-test design, should be viewed as just one of the several kinds of change upon which the efficacy of the Project must be judged.

The following pages provide a detailed account of the development of the Sound topic, the Project findings and the implications which they raise for science education.

1. METHODOLOGY

1.1 Sample

a. Schools

Six schools belonging to Lancashire LEA, five primary and one junior school, were involved in the 'Sound' topic. Two teachers and their classes were involved from five of the schools and one teacher from the sixth school. At least one class from each school year group was included, from reception to fourth year junior, so the whole primary age range was covered in the sample. Names of the participating schools, teachers and head teachers may be found in Appendix I.

b. Teachers

The teachers were selected to participate in the Project by their LEA. Neither a background knowledge of science nor an expertise in teaching science were pre-requisites for involvement. The teachers had already been working with the Project for a year and had had the opportunity to develop an awareness of project methodology and techniques as a result of working on the topic of 'Growth'.

c. Children

All the children in the classes of participating teachers were involved in the Project work to some extent, particularly in those phases which were classroom-based, the Exploration and the Intervention. A smaller, random sample from each class was also selected to be interviewed individually about their ideas. This sample was stratified by age, gender and achievement. This third variable was based on the teachers' subjective judgements of children's overall scholastic achievement as high, middle or low.

d. Liaison

The teachers were supported by a member of the Lancashire advisory teacher team. This advisory teacher, by virtue of being closely involved with six Project schools, was able to provide support and encouragement for the teachers and to liaise closely with the University research team. The advisory teacher also carried out some interviews with individual children.

1.2 The Research Programme

Classroom work concerned with the topic of 'Sound' took place during the second and third terms of the academic year 1987-88. Each classroom phase was followed by a period of one-to-one interviewing by the Project team. The timetable was as follows:

Exploration (March 1988)
  Pre-Intervention interviews (late March 1988)
  Intervention (May 1988)
  Post-Intervention interviews (June 1988)

1.3 Defining 'Sound'

A list of concepts concerning sound was compiled to provide a framework for work in this area.

1. Sound is produced by a vibrating object (something moving rapidly to and fro).

2. Sound travels as vibrations through a medium.

3. Sound travels through some media better than others.

4. The vibrations are picked up by a specialised receiver: in land mammals, they are received by the ear. The ear transmits sensory information to the brain, which translates it as 'sound'.

5. Vibrations can also be picked up by other senses e.g. touch, sight.

6. An object gives a natural note when it vibrates. This note can be changed.

This list was for the use of the Project team. It was not intended to be presented to teachers as a basis for their classroom work.

2. *PRE-INTERVENTION ELICITATION WORK*

Prior to the Exploration phase, the teachers and Project team were involved in a total of two-and-a-half days of Project meetings. These meetings built upon the experiences the teachers had of SPACE work and enabled them to participate in discussions leading to the development of the 'Sound' Exploration activities. Additionally, teachers were asked to consider the criteria which would be important for the Exploration activities to be successful in the classroom. These criteria may be found in Appendix II.

2.1 Exploration (March 1988)

Teachers were given a pack which contained descriptions of the activities which they were asked to use with their classes. Because of the preparatory work the teachers had done at the meetings, it was possible to make these descriptions brief, containing only the amount of instruction necessary to ensure uniformity of presentation between classes. This pack may be found in Appendix III.

Activities

Teachers were asked to set up the following activities in their classrooms for a period of four weeks and to encourage children to interact with them while not teaching the class anything about them.

a. sending and receiving a message through a string telephone

b. stretching and plucking a rubber band

c. listening to sounds through an ear trumpet

d. hitting a drum which had some rice grains on the skin

e. listening to everyday sounds

Each activity is described in detail below, accompanied by an indication of the reactions of pupils and teachers to the activity.

a. Sending and receiving a message through a string telephone

A string telephone was constructed from two yogurt cartons, each with a small hole in the base, and a length of string. The string was threaded through the holes, into the cups, and knotted securely at each end. Thus, the string provided a link between the two yogurt cartons.

The children were asked to work in pairs and to hold one yogurt pot each and take it in turns to whisper a message to each other. It was suggested that younger children should make their message an instruction, e.g. draw a house, so the drawing would show whether the message had been heard correctly. Older children could, as an alternative, write down what had been said to them. The string telephone was chosen because it made explicit a pathway for sound travel between the speaker and receiver, thus exemplifying sound transmission through a medium. This activity was very popular with the children and encouraged them to express their ideas about sound travel, though several teachers remarked that the children had not known how to use the apparatus correctly: some children had tried to bend the string round a corner, while others had wanted to face their partners and had thus pressed the string against the bottom of one of the yogurt pots. Each of these arrangements would prevent the string from vibrating freely and would hamper the hearing of the message.

b. Stretching and plucking a rubber band

Children were asked to stretch a rubber band either around a carton or between their fingers and to pluck it to see whether they could play a tune. This activity was chosen to exemplify sound production and the movement associated with it. The movement could be both seen and felt as a tingling sensation in the fingers which were stretching the band. This activity produced some useful information about links which children made between movement and sound production, and also between the tension in the band and the pitch of the resultant sound. Some teachers were initially apprehensive about the potential misuse of the bands by the children but these anxieties were not confirmed.

c. Listening to sounds through an ear trumpet

The top half of a large plastic lemonade bottle was used as an ear trumpet, or 'funnel', to encourage children to focus on sound reception. Children were asked to put the funnel to one ear and to listen to the sounds around the classroom. They were then asked what difference the funnel made to the way they could hear. Some children chose to use the funnel with the wide opening towards their ear rather than away from it and this probably accentuated the effect which was noted by many children, that the funnel acted like a sea-shell: pupils often claimed that the funnel made them hear differently, that they could hear a noise like the sea. This effect might have been achieved partly because of the instruction to listen to sounds around in the classroom rather than to focus on one particular quiet sound, such as a watch or clock ticking. A different instruction might have helped children to focus more easily on the perceived amplification of the sound rather than other peripheral effects. The peripheral, distorting effects might have been exacerbated by the length of the lemonade bottle section which was used. The long length tended to incorporate an echo chamber which could have obsured the sound focusing effects. Despite this draw-back, the funnel was a very useful activity for encouraging children to attempt to represent sound diagrammatically and for suggesting mechanisms concerned with how sound enters the ear.

d. Hitting a drum which had some rice grains on the skin

Children were asked to hit a drum both before and after putting rice grains onto its surface. They were also asked to put their finger lightly onto the drum's surface while it was making a sound. This activity primarily exemplified sound production, but the children were also asked to comment on transmission and reception. Teachers reported that the drum was very popular with the children, though rather noisy within the classroom. The beating of the drum encouraged children to express a wide range of ideas about sound production and it was a valuable activity. However, the presence of the rice was a distraction for all except the oldest children who could explain the movement of the rice from their ideas about how the drum made a sound.

e. Listening to everyday sounds

Teachers were asked to take their children on a 'listening walk' around the school and the grounds, to focus their attention on sounds. This was a productive introductory activity to the work on sound but did not itself lead to the expression of very many ideas.

Elicitation Techniques

Teachers were asked to refrain from class discussion so that each child's ideas would not be influenced by their peers more than could be avoided. As the teachers were being careful not to teach the children about sound it was suggested that, as well as asking individual children to draw their ideas, there should be a class log-book into which entries could be made informally by children. This book would provide teachers with a way of showing an interest in what children were doing and in their ideas.

Class Log-book

It was suggested that teachers introduced a 'Sounds' or 'Listening' book into which children could write or draw their ideas and observations about any of the Exploration activities. It was also suggested that children might like to make entries concerning types of sounds they had heard, and also different sound makers which they had experienced recently.

In practice, many teachers chose either to use a separate book for each activity, or to leave some paper by each activity so that entries could later be compiled into one book. This book provided an informal record for teachers of the children's thoughts about the topic. The notion of a 'Sounds' book tended to be identified with children's reactions to their 'listening walk'.

Labelled Diagrams

The Project team had specified two activities for which children should be asked to draw their ideas. These were the drum and the ear trumpet. The drawings which children were asked to produce were intended to portray ideas rather than be aesthetically pleasing pictures. This distinction had initially needed careful reinforcement by teachers, since such diagrams were not a form of recording with which many children were familiar. The questions asked were:

a. Drum

Draw pictures to show:

i. how you think the drum makes a sound;

ii. how you think the sound gets from the drum to you so you can hear it;

iii. how you think you hear the sound.

b. Funnel ear trumpet

Draw how the funnel helps you to hear differently.

These questions were very productive in helping children to express their ideas.With hindsight, an additional instruction to show the sound on the picture might have encouraged more children to attempt to represent their conception of sound on paper. Teacher annotation of the diagrams, following discussion with the child, was invaluable in clarifying the children's ideas.

One interesting development in the teacher's perception of their role during this, the second topic with which they had used Project techniques, appeared to be an increasing sense of ownership of the techniques and a desire to find out for themselves the ideas which were held by their classes. This development resulted in some teachers posing their own questions to their classes and asking them to draw additional diagrams to explain the rubber band and the string telephone. One teacher, for example, asked children to draw how they thought they heard the sound from the rubber band. This instruction was very similar to that used by the Project team in connection with the drum. While it was carefully worded, it might not have been the most profitable instruction to use with the rubber band, where the most salient aspect was sound production. The majority of children's drawings appeared to be responses to a more general instruction to draw how the rubber band made a sound.

Individual Discussion

Teachers found the diagrams to be a profitable starting point for individual discussion and they were keen to question the children as much as was practicable in the classroom. Increased confidence with open questioning methods made this activity more productive and enjoyable for both teachers and pupils. This questioning, in conjunction with the other elicitation techniques, enabled pupils to clarify their ideas and to express them to the teachers in a coherent manner. Both pupils and teachers had previously worked on the 'Growth' topic together and it is likely that this experience had facilitated the participation of both parties in these discussions. Teachers were likely to have become more skilled in asking questions of a kind which offered children the opportunity to express their ideas; the children could have improved their self-expression and been more familiar with the more open role of the teacher.

Interviews with Individual Children

A sample consisting of approximately six children from each of the eleven classes was interviewed individually about each of the Exploration activities. This sample was balanced for gender, achievement and age and was selected randomly within these constraints. Each of the children interviewed had had access to the Exploration activities for at least two weeks. Members of the Project team visited the children's schools to interview and talked to children either in a quiet corner of the classroom or in an empty room, for example the library or staff room, within the school building. The interviews were conducted in an informal manner so that the children were as much at their ease as possible. The interviewers had a list of questions demonstrating the areas in which there should be an attempt to elicit children's ideas. In order to maintain informality and to supply further information these questions could be supplemented with additional ones where necessary, or re-phrased where children were unclear what was being asked. An important consideration throughout the interviews was that children should not feel pressurised into giving answers which were not their considered opinions. With this in mind, 'I don't know' was accepted as a valid response.

The length of interview varied greatly from one child to another but the modal duration was 30-45 minutes.

The questions around which the interviews were centred may be found in Appendix IV.

Organisation

The Exploration activities required minimal apparatus and so were easy to set up in the classroom. The main drawbacks, when dealing with the topic of 'Sound', were that generating sound made a noise, and that careful listening to the sound required low levels of background noise. These constraints sometimes led to work being carried out during playtimes when it was not possible to place the activities away from the main class yet in a position where the teacher could supervise the children who were involved. Some teachers found that certain activities were best used with the whole class together. Examples of suitable whole-class activities were the 'listening walk' and the rubber bands, both of which were quiet rather than noisy activities.

Summary

The Exploration activities were each very different in the experiences which they provided, but each of them was profitable in terms of encouraging children to express their ideas concerning particular aspects of sound. The activities were:

a. sending and receiving a message through a string telephone (sound transmission)

b. stretching and plucking a rubber band (sound production)

c. listening to sounds through an ear trumpet (sound reception)

d. hitting a drum which had some rice grains on the skin (sound production)

e. listening to everyday sounds.

The pupils and teachers were both more familiar with expressing ideas and encouraging their expression due to earlier involvement with the SPACE Project. This familiarity enabled teachers to feel confident about their class's ideas and to develop a picture of possible avenues which the children could profitably investigate during the Intervention phase in their classrooms.

3. CHILDREN'S IDEAS

Part 1: An informal look at children's ideas

The children in each of the classes participating in Project work were asked by their teachers to record their ideas about sound. Their recorded ideas were responses to questions which were in some cases suggested by the Project team and in other cases posed by the teachers themselves. The teachers of infants accumulated most of their information in the form of notes made during discussions with pupils while the teachers of juniors made more use of individual diagrams. The ideas contained in these diagrams have been categorised in a manner compatible with the analysis of the interviews, and frequency counts have been performed. Where percentage figures are given the sample size is stated alongside them.

The children's drawing and writing provided a very rich source of data regarding children's ideas, and they allowed access to aspects of the topic which were difficult to probe in individual interview. For example, some insight was provided into children's representations of sound, and into their perceptions of the practical activities with which they were involved.

This section considers children's ideas about sound under the following headings:

3.1.1 *What is sound?*

a. the perceived association between sound vibrations.
b. the causal relationship between sound production and vibrations.
c. the recognition of attentional effects in hearing.

3.1.2 *The production of sound*

a. Explanations in terms of physical attributes and conventional usage of objects.
b. Explanations in terms of the application of a force.
c. Locations and mechanisms for sound production: the role of vibrations.
d. The development of a generalized concept of sound production.

3.1.3 *The transmission of sound*

a. Explanations in terms of attributes of the listener and the sound producer.
b. Sound 'travels'.
c. Sound travel in the absence of media.
d. Sound travel through string.
e. Sound travel through air.

3.1.4 The reception of sound

a. Sound reception by the ear.
b. The funnel as a sound box.
c. The effects of pressure and compression of sound.

3.1.5 Children's representations of sound

3.1.6 Vocabulary used in association with sound

3.1.7 Summary

3.1.1 What is sound?

a. The perceived association between sound and vibrations

Sound is something which cannot be seen directly but which can be experienced through vibrations in the media of hearing, seeing and touching. It is a phenomenon which is accessible to every child through at least one sensory modality. Children can have experiences both of producing sounds and of receiving them since sound is produced, transmitted and received as vibrations through particular media. There is an inextricable link between sound and vibrations which is there for children to observe and incorporate into their constructs, and children made different degrees of association between sound and vibration.

Many young children did not suggest any association between movement or vibration in the sound producer.

Fig. 3.1

(Age 6 years)

"The drum stick makes the sound."

The example given in Fig. 3.1 shows how the child considered the action of the drumstick hitting the drum to be sufficient to generate the sound. The rice which was on the skin of the drum and could have led children to observe the skin vibrating was either ignored by children of this age or was itself made the focus of attention, implicated in sound production to the exclusion of the stick or drum.

Fig. 3.2

(Age 7 years)

"The rice is going up and down to make the sound."

The plucking of a rubber band provided an interesting example of perceptible vibrations and half of the infant and lower junior children (n = 54) observed and commented on the movements of the band. These comments were not, however, related to sound production.

Fig. 3.3

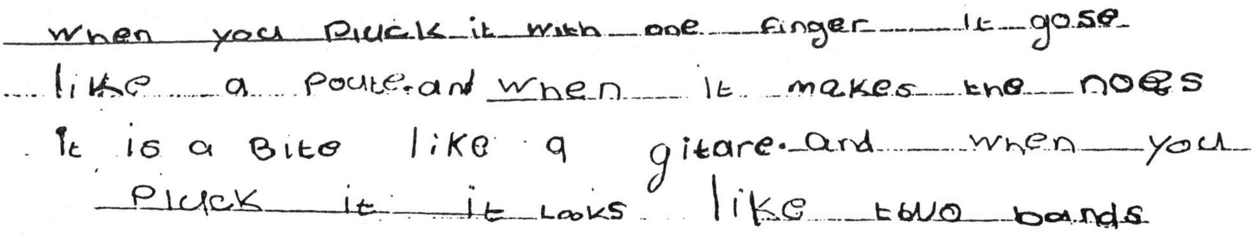

When you Pluck it with one finger It gose
like a poute and when It makes the noes
It is a Bite like a gitare and when you
Pluck it it Looks like two bands

(Age 6 years)

"When you pluck it with one finger it goes like a ? and when it makes the noise it is a bit like a guitar and when you pluck it it looks like two bands."

It is possible that these children had not yet had wide enough experiences of vibration in the context of sound production for them to find it helpful to associate the two events. Fig. 3.4 is an example of a child who had started to form such an association.

Fig. 3.4

(Age 7 years)

"It makes a tune because the box is long. When you pull it back (elastic band) it shivers. When it stops making a noise it doesn't move."
(Teacher's note)

b. The causal relationship between sound production and vibrations

Children who had made an association between sound production and vibration did so in several ways, and it seems likely that the context of the observations might have influenced the link that was made. Links between sound and vibration were often suggested in terms of cause and effect, so that sound was either caused by vibration, or vibration was caused by sound. A small number of children thought that vibrations and sound were the same thing. Very few responses linking sound and vibration were made by children in the two younger age groups, infants and lower juniors.

i. Vibrations cause sound

Of the range of activities with which children interacted, the rubber band provided the context which generated the most mention of sound being caused by vibrations (32% of upper juniors, n = 31) and this link was made by more girls than boys (47% of girls, 14% of boys).

Fig. 3.5

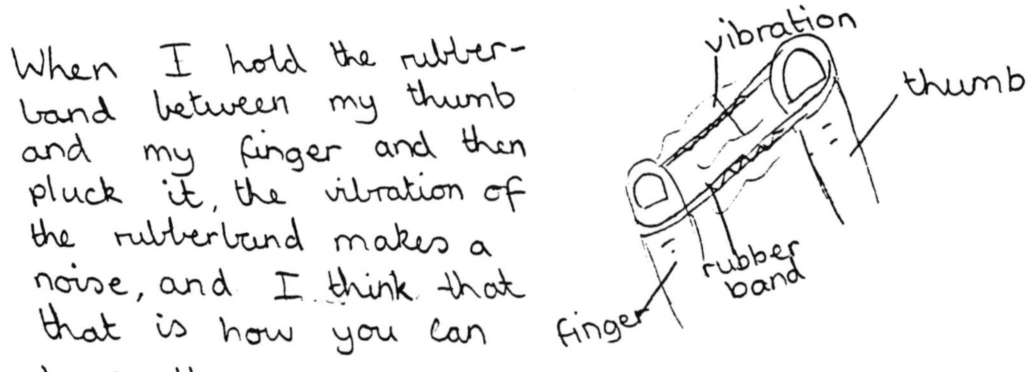

When I hold the rubber-band between my thumb and my finger and then pluck it, the vibration of the rubberband makes a noise, and I think that that is how you can hear them.

vibration
thumb
rubber band
finger

(Age 9 years)

"When I hold the rubber-band betwen my thumb and my finger and then pluck it, the vibration of the rubber band makes a noise, and I think that that is how you can hear them."

Children were directly involved in generating movement in the band by stretching and releasing it so this might have been another factor, together with the perceptibility of the vibrations, which encouraged children to mention this idea. Children's ideas concerning the drum also tended to suggest that vibrations caused sound, though some children did suggest that sound caused vibrations.

Fig. 3.6

(Age 10 years)

"When I beat the drum it sort of trembles and as the drum trembles it makes a sound."

Sound causing vibrations was the more common explanation given for the string telephone and the vibrations seemed to be associated with the sound travelling rather than the sound being produced. This could have been due to the nature of the activity, which exemplified sound transmission and children might not have focused their attention on the production of speech.

Fig. 3.7

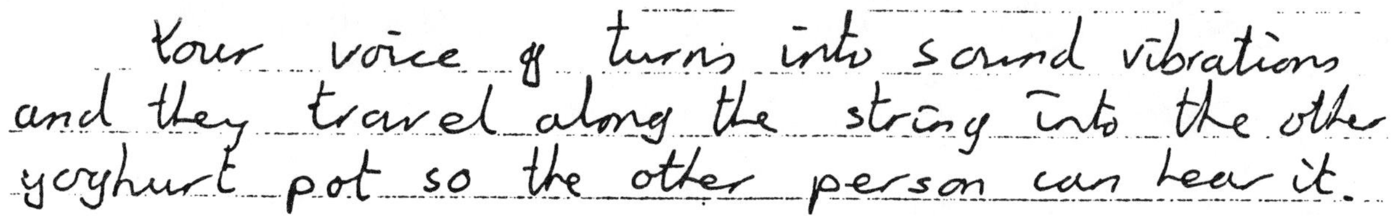

Your voice turns into sound vibrations and they travel along the string into the other yoghurt pot so the other person can hear it.

(Age 9 years)

"Your voice turns into sound vibrations and they travel along the string into the other yoghurt pot so the other person can hear it."

ii. Sound is vibrations

Some children's writing and drawing seemed clearly to suggest that sound and vibrations could not be separated.

Fig. 3.8

ALastic Band
It makes a nose that goes "ping" and vibrates will it pinging. And the sound cares on will it's vibrating.

(Age 7 years)

"Elastic band. It makes a noise that goes 'ping' and vibrates while it's pinging. And the sound carries on while it's vibrating."

Fig 3.9

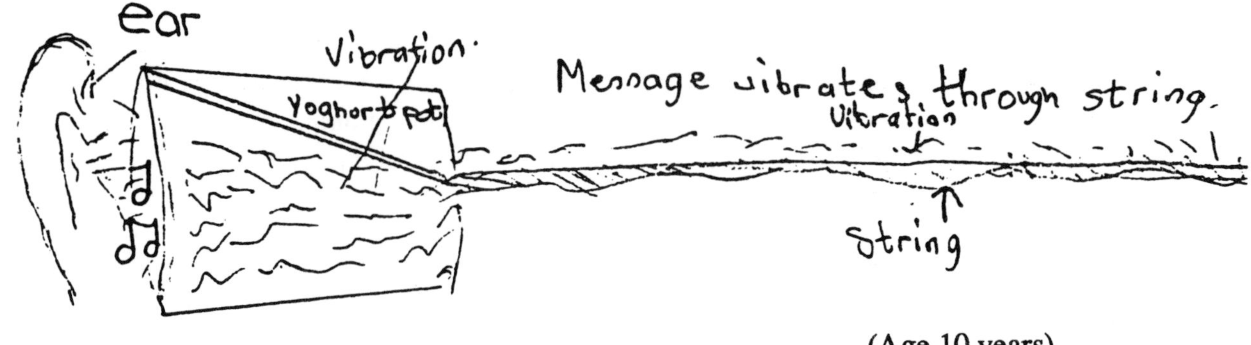

(Age 10 years)

c. *The recognition of attentional effects in hearing*

There is an etymological distinction between hearing and listening, and it is a distinction which a small number of children in each age group emphasised in their answers to questions about how they heard: they asserted that they heard because they were listening. This emphasis on the need for the hearer to be attending to the sound was sometimes given as a complete explanation for hearing (Fig 3.9). However, it was often an accurate and valid comment as a component part of a more complex response (Figs. 3.10, 3.11). The psychological model of 'active listening' shows an awareness of the effects that concentration and attention play in hearing.

Fig. 3.10

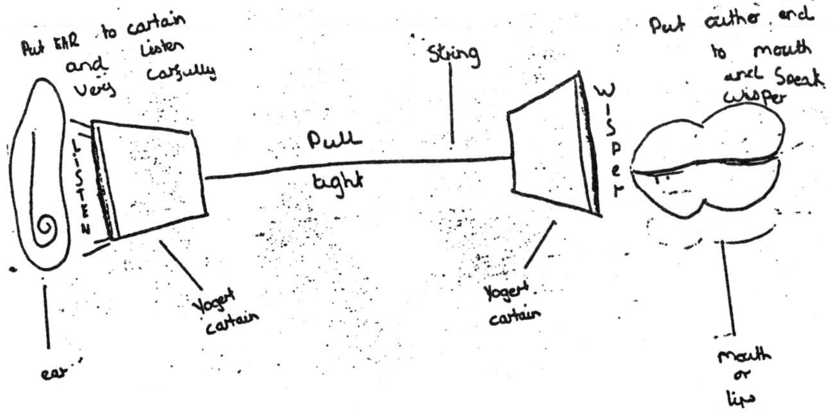

(Age 9 years)

Fig. 3.11

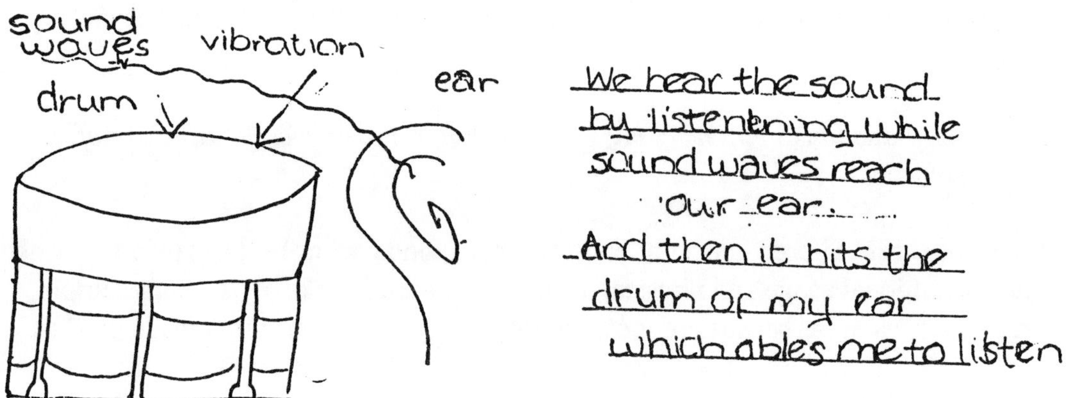

(Age 10 years)

"We hear the sound by listening while sound waves reach our ear. And then it hits the drum of my ear which ables me to listen."

Fig. 3.12

(Age 7 years)

"The noise goes through the air to people who are listening. If they're not listening they'll only hear a bit of it and not all of the sound."
(Teacher's note)

3.1.2 The Production of Sound

In order for a sound to be produced by an object, the object must vibrate. Vibrations are caused by an input of energy, often a physical action which results in an observable impact. There are, therefore, many contingent variables related to sound production which can be observed by children. Some involve gross features of the system such as the impact, the implement, the action delivering the hit or physical attributes of the sound producer. Other observations may be of less obvious characteristics, such as vibrations. The ideas which a child has about sound production are likely to relate to the range and quality of their experiences and observations.

a. Explanations in terms of physical attributes and conventional usage of objects

Many younger children appear to consider that sounds occur because they know about the conventional usage and physical attributes of particular objects. For example, a drum makes a noise because of the stick (Fig. 3.13), or because it is made of plastic.

Fig. 3.13

The drum made the sound
and the stick the drum noise

(Age 6 years)

"The drum made the sound and the stick the drum noise."

Fig. 3.14

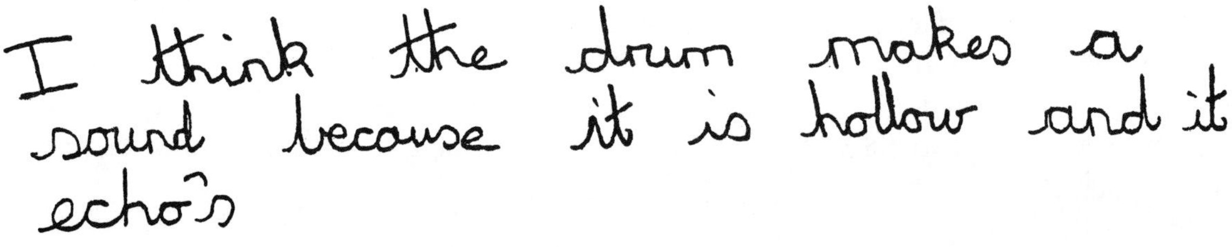

I think the drum makes a sound because it is hollow and it echo's

(Age 7 years)

"I think the drum makes a sound because it is hollow and it echo's."

Fig. 3.15

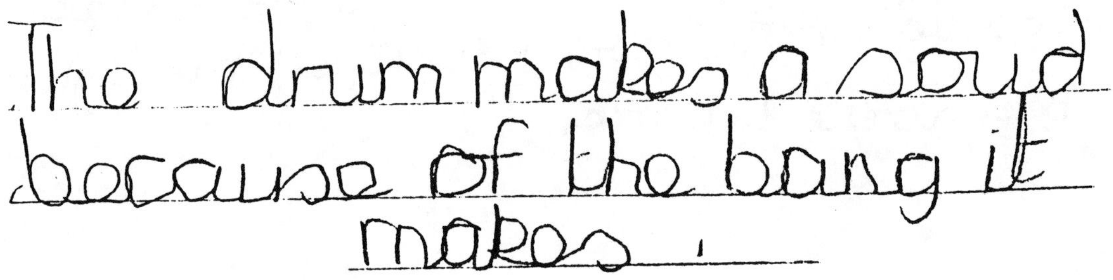
The drum makes a soud because of the bang it makes.

(Age 7 years)

"The drum makes a sound becuase of the bang it makes."

Children had similar ideas about the rubber band which was thought to make a sound because of the composition of the band and properties of the rubber (Fig. 3.16).

Fig. 3.16

"It's because it's thin elastic. It flicks up and down when you let it go."
(Teacher's note)

(Age 7 years)

The degree of tension in the rubber band or the drum head was also related to the pitch of the note produced. The number of children expressing this idea increased with age to approximately one quarter of the upper junior sample (n = 31) for the rubber band. In the rubber band activity children were asked whether they could play a tune on the band, and this might have encouraged more children to attend to the variables of tension and pitch. Fig. 3.17 shows the only child to have mentioned the word 'pitch'.

Fig. 3.17

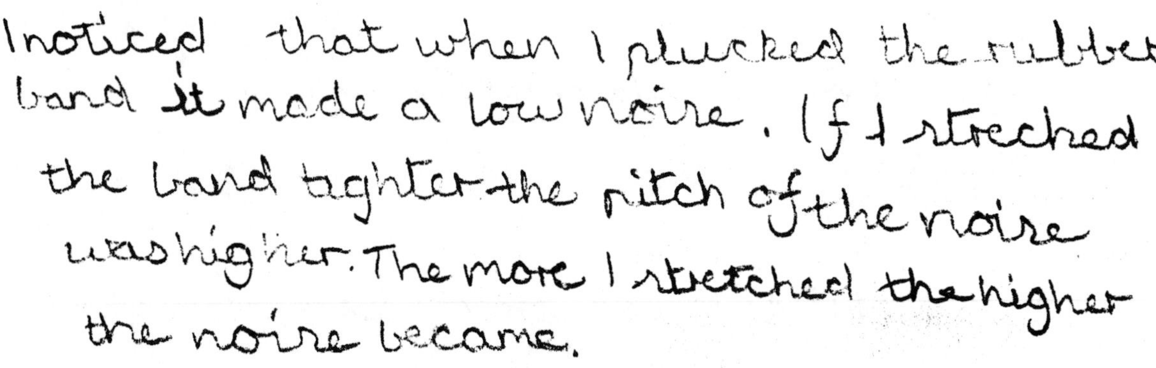

I noticed that when I plucked the rubber band it made a low noise. If I streched the band tighter the pitch of the noise was higher. The more I stretched the higher the noise became.

(Age 9 years)

"I noticed that when I plucked the rubber band it made a low noise. If I stretched the band tighter the pitch of the noise was higher. The more I stretched the higher the noise became."

b. *Explanations in terms of the application of a force*

As illustrated by Figs. 3.18 and 3.19, the force used to generate the sound from the instrument was often commented upon. Such explanations made an attempt to suggest a mechanism of sound production. The statements were related to observations which the children had made.

Fig. 3.18

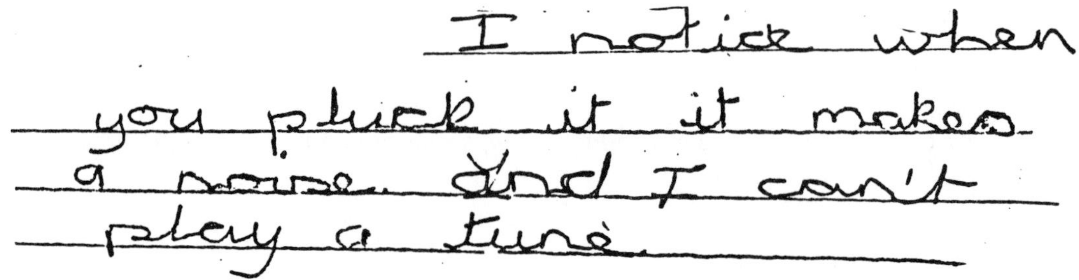

(Age 7 years)

"I notice when you pluck it it makes a noise and I can't play a tune."

Fig. 3.19

The bibrisions
OF the impact
makes the
drum sound

powor of
the impact

(Age 9 years)

"The vibrations of the impact makes the drum sound."

c. Locations and mechanisms for sound production

Mechanisms by which a drum produces sound were suggested by a number of children. The proportion of children suggesting a mechanism increased with age from 13% of infants (n = 24) to 98% of upper juniors (n = 84). The mechanisms involved two possible locations for sound generation: inside the drum and at the surface of the drum.

A variety of mechanisms of sound production was mentioned to explain the process of making sound at each location.

i. Sound generation inside the drum

Of the children suggesting a mechanism, inside the drum was the location suggested by all of the infants and just over one third of juniors. The most common ideas were that the sound inside the drum 'echoed around' (Fig. 3.20) or that the inside of the drum vibrated. Alternatively, the inside of the drum was thought to make a sound when it was compressed by the impact lowering the drum head (Figs. 3.21 and 3.22). The existence of air inside the drum was mentioned by some upper junior children (Fig. 3.22). It is unclear whether the air was thought to fill the inside of the drum, or whether it was a convenient label for what was really thought to be an empty space.

Fig. 3.20

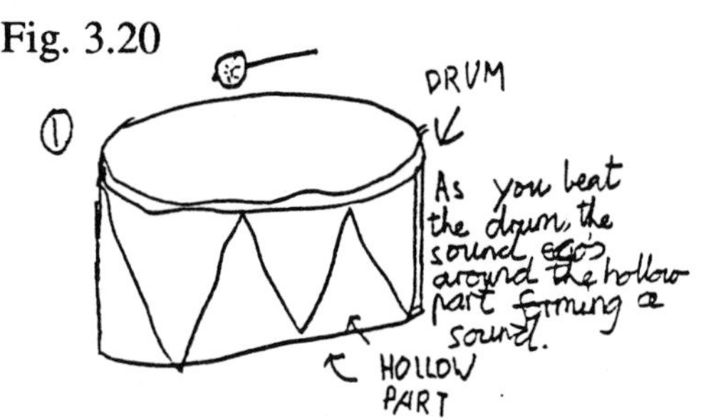

"As you beat the drum, the sound echoes around the hollow part forming a sound."

"As you beat the drum sound echoes like I said and then comes up and seeps through the skin on top."

"As the sound echoes out of the drum it forms a deep sound for you to hear".

(Age 10 years)

Fig. 3.21

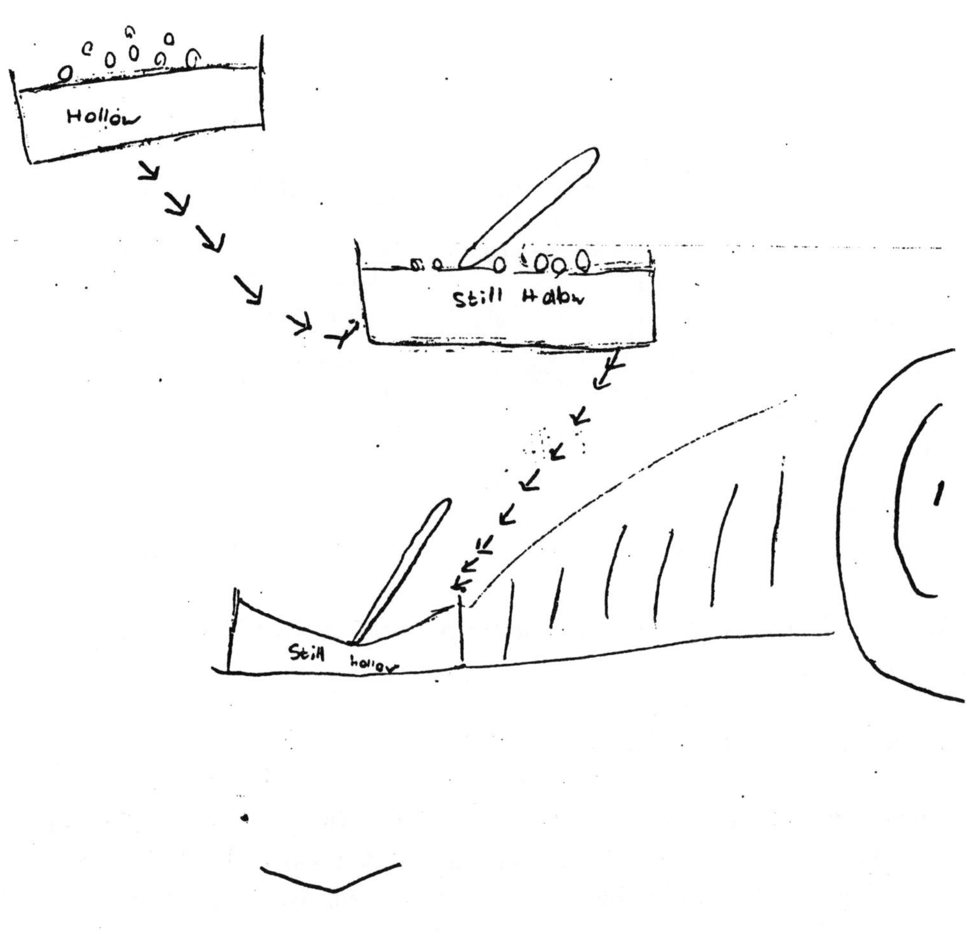

when it goes in we here
the vibration

(Age 9 years)

"When it goes in we hear the vibration."

Fig. 3.22

(Age 10 years)

"I think that when you beat the drum the air pushes down and I think that makes the sound."

ii. Sound generation at the surface of the drum

The surface of the drum as the location of sound production was suggested by two thirds of juniors. The most common mechanism was that the drum head vibrated (Fig. 3.23). This explanation, while accurate, was only rarely accompanied by any mention of any other vibrations, e.g. sound waves.

Fig. 3.23

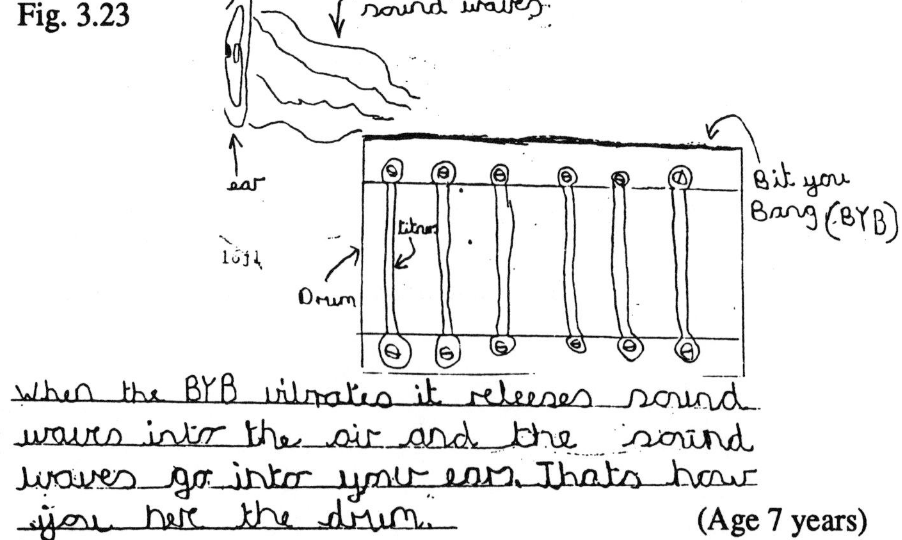

(Age 7 years)

"When the BYB vibrates it releases sound waves into the air and the sound waves go into your ears. That's how you hear the drum."

Some children suggested that the air inside the drum caused the drum head to vibrate (Fig. 3.24).

Fig. 3.24

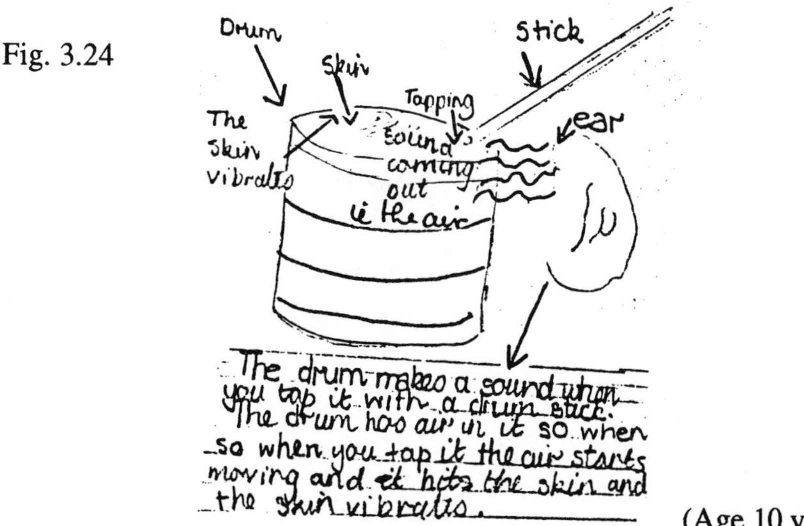

(Age 10 years)

"The drum makes a sound when you tap it with a drum stick. The drum has air in it so when you tap it the air starts moving and it hits the skin and the skin vibrates."
(Some teacher annotation)

Other children suggested that the sound came straight off the drum surface. Fig. 3.25 might be an attempt to articulate the fact that the head vibrated, though Fig. 3.26 appears to contain no suggestion that the drum head did anything other than deflect the sound upward after the blow from the stick.

Fig. 3.25

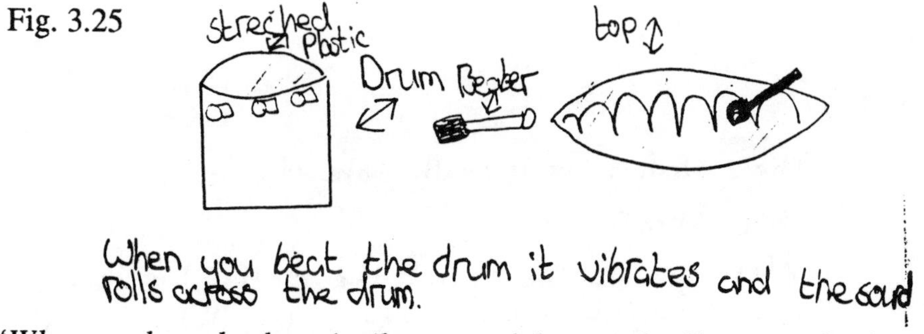

"When you beat the drum it vibrates and the sound rolls across the drum."

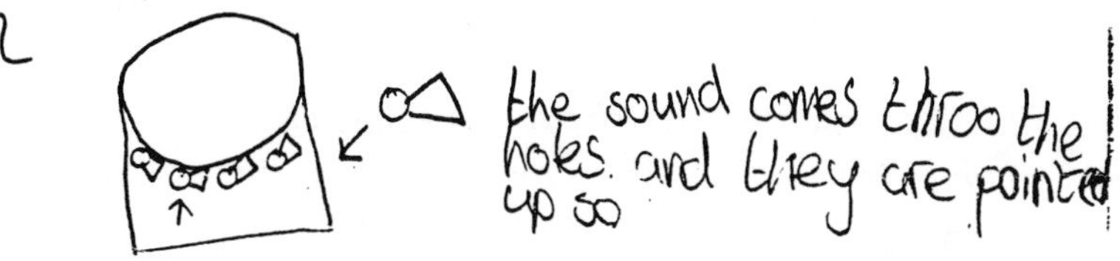

(Age 10 years)

"The sound comes through the holes and they are pointed up so."

Fig. 3.26

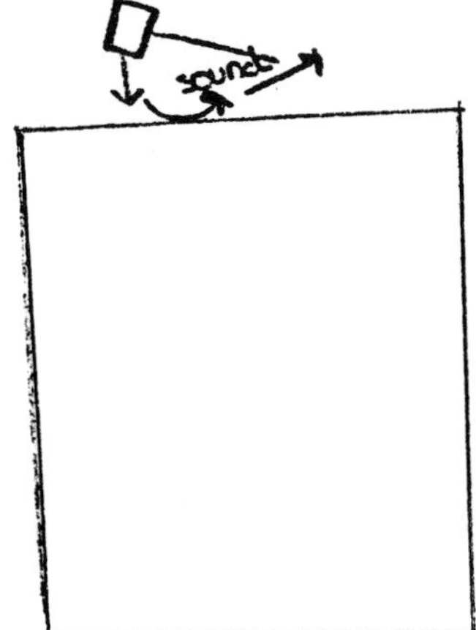

The sound is bounced off the cover when you hit it.

(Age 10 years)

"The sound is bounced off the cover when you hit it."

iii. Sound generation involving the surface and the inside of the drum

A mechanism was described in which the sound appeared to start at the surface of the drum and then to move downwards through the drum until it reached the bottom. It then either came out of the bottom of the drum (Fig. 3.27) or was deflected back up and out of the top (Fig. 3.28). A further, connected elaboration suggested that the drum head was made to vibrate by the sound as it passed from the inside of the drum, out through the head (Fig. 3.29).

Fig. 3.27

The strings underneath make the sound. –
they vibrate
When you hit the top of the drum the noise
goes through the drum and makes strings vibrate
There is nothing in the drum

(Age 7 years)

"The strings underneath make the sound - they vibrate.
When you hit the top of the drum the noise goes through the drum and makes strings vibrate.
There is nothing in the drum."
(Teacher annotation)

Fig. 3.28

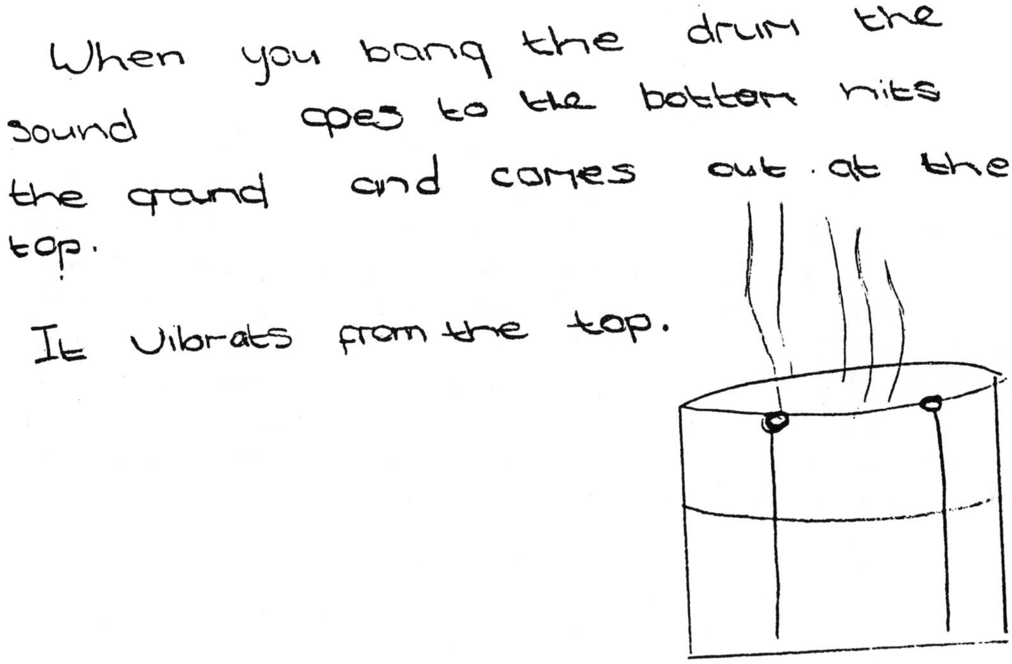

(Age 10 years)

"When you bang the drum the sound goes to the bottom, hits the ground and comes out at the top. It vibrates from the top."

Fig. 3.29

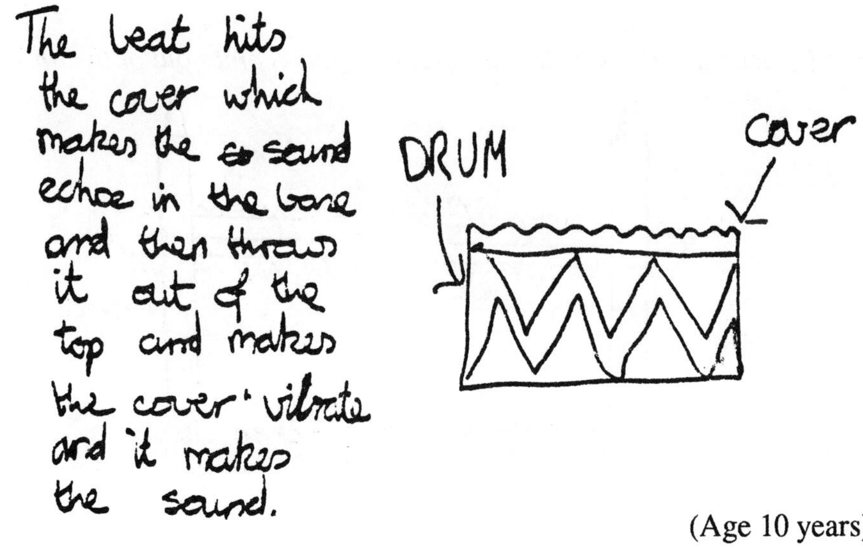

(Age 10 years)

"The beat hits the cover which makes the sound echo in the base and then throws it out of the top and makes the cover vibrate and it makes the sound."

The mechanisms suggested in Figs. 3.27 to 3.29 could be considered to have much in common with a conventional explanation for sound production. However, air was not mentioned and it is possible that the sound was regarded as a discrete entity rather than as vibrations travelling through a medium.

There are few obvious parallels between the mechanisms suggested for the production of sound from a drum and those from a rubber band, possibly because the rubber band does not have a built-in resonating chamber. The similar mechanisms which were found were given by children who stretched their elastic bands round a box before plucking it (Fig. 3.30). The rubber band which was plucked without using a box generated only a small range of ideas, all related to vibrations travelling round the band (Fig. 3.31). The vibrations were sometimes suggested to turn into sound when they came into contact with something, for example the pencil or the thumb, which was keeping the band stretched. Children seem to have used observations made in a range of sense modalities to influence their ideas.

Fig. 3.30

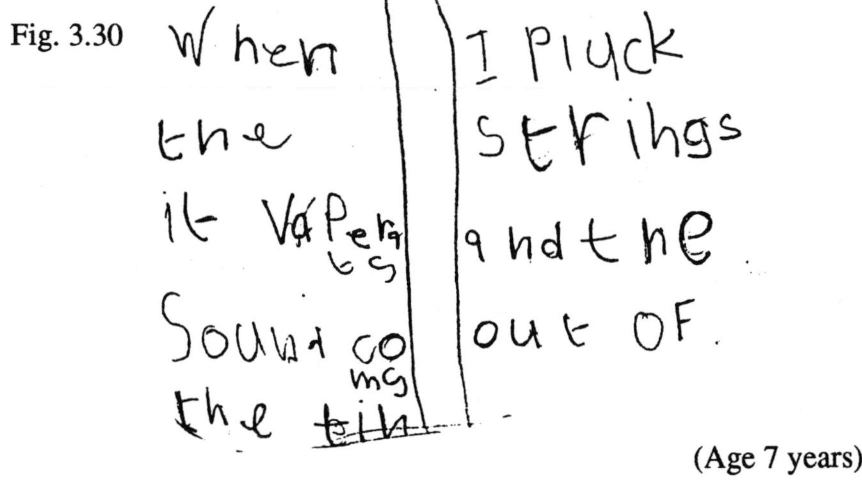

(Age 7 years)

"When I pluck the strings it vibrates and the sound comes out of the tin."

Fig. 3.31

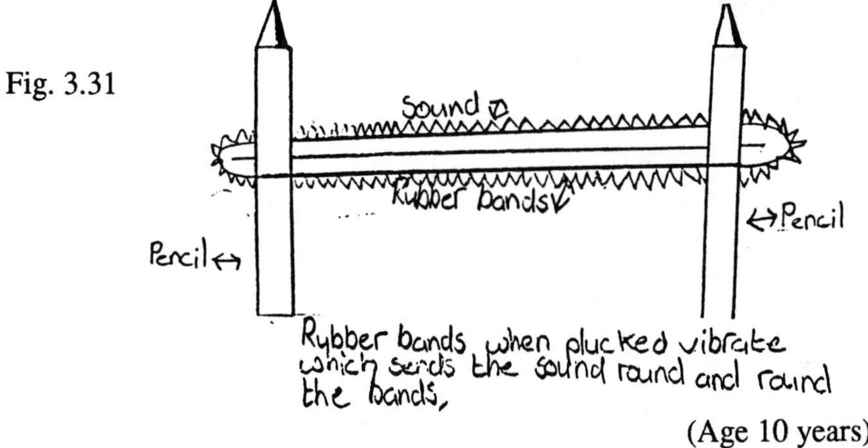

(Age 10 years)

"Rubber bands when plucked vibrate which sends the sound round and round the bands."

The specific mechanisms which children suggested for sound generation appeared to be context-specific, with the rubber band responses being different from the drum. In principle, though, there were common groups of response, making reference to physical attributes of the object, to the force needed to generate sound or to vibrations.

d. The development of a generalised concept of sound production

Fig. 3.32 shows how one child has made a generalized observation from his experiences of the production of sound.

Fig. 3.32

Nearly every thing you hit makes
a Sound

(Age 7 years)

"Nearly every thing you hit makes a sound."

This statement is interesting since it contains a generalization which suggests that objects can be made to make a sound by hitting them. This generalization between experiences was not very common and more children focused upon the physical attribute of a specific object when offering an explanation for sound production.

Fig. 3.33

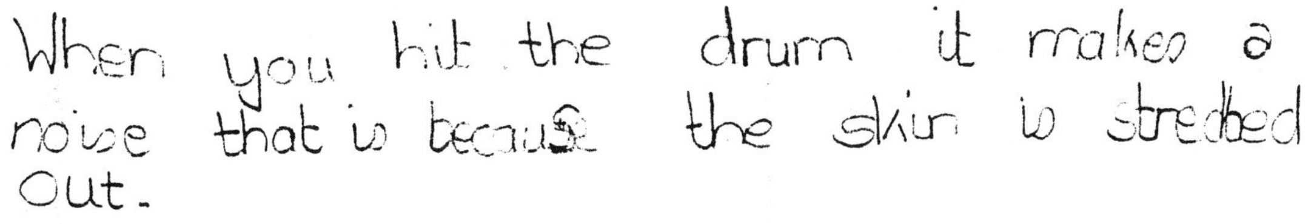

When you hit the drum it makes a
noise that is because the skin is stretched
out.

(Age 10 years)

"When you hit the drum it makes a noise that is because the skin is stretched out."

Fig. 3.32 and Fig. 3.33 show contrasting approaches to the problem of understanding why objects make sounds. In Fig. 3.32 the child has observed that there is a relationship between hitting things and their making sounds. There is no evidence of any ideas about why there might be exceptions within this relationship, for example related to the tension in the object. The child whose work is represented in Fig. 3.33, on the other hand, has observed the need, in a specific case, for tension to be present in order to make a sound. There is no evidence that this child has generalised this necessary condition to other sound makers. It would be possible to speculate about whether starting from a generalised statement and determining the exceptions to the rule, or generalising from a specific example would be more profitable than the other in terms of the conceptual development which could be promoted.

3.1.3 Transmission of Sound

a. Attributes of the listener and the sound producer

In describing how hearing is possible, many younger children referred both to attributes of themselves as listeners and to attributes of the sound producers without mentioning sound travelling. These attributes were:

. proximity
. volume
. attentional factors
. attributes of the Exploration activity

Fig. 3.34 for example, shows dogs barking across a road. The child whose drawing this was said they could hear the dog because of its proximity and the volume of the bark.

Fig. 3.34

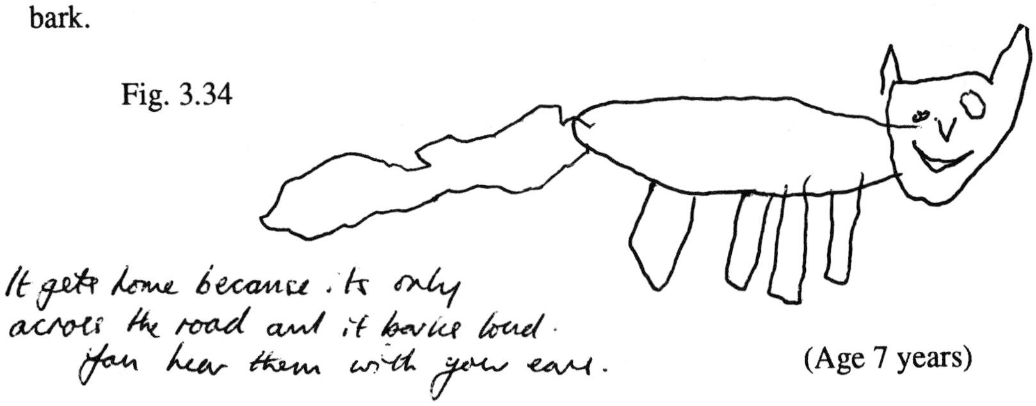

(Age 7 years)

"It gets home because it's only across the road and it barks loud. You hear them with your ears." *(Teacher annotation)*

The range of playground noises in Fig. 3.35 was thought to be heard by means of ears, because ears listen all the time. This response is another example of 'active listening', the particular psychological perspective on hearing which seems to be held by certain children.

Fig. 3.35

(Age 7 years)

"In the playground. It gets to you because you listen. Yes it would get to you if you didn't listen because your ears don't stop listening. They're listening all the time even at night." (Teacher annotation)

In order to suggest any of these explanations children needed to make observations, all of which would be relevant to the notion of sound travelling and would not be inconsistent with an explanation that incorporated transmission. However, these children have not extended their ideas to include an explicit mention of sound travelling.

b. Sound 'travels'

A large number of children did consider sound to travel, and a variety of pathways through which the sound could travel were mentioned. The explicit suggestion that sound went from one place to another was sometimes not elaborated further, as shown in Fig.3.36

Fig. 3.36

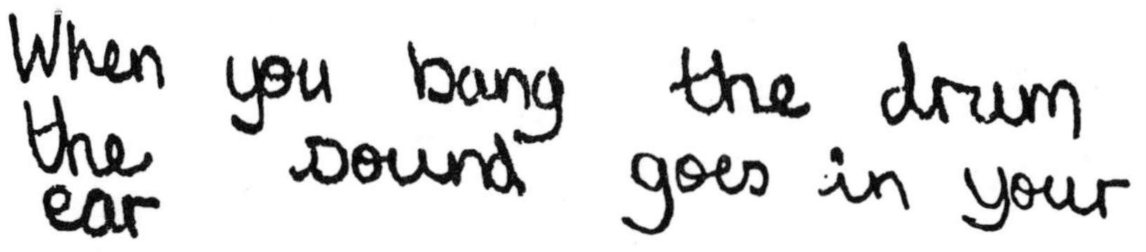

When you bang the drum
the sound goes in your
ear

(Age 7 years)

"When you bang the drum the sound goes in your ear."

This response was often linked with the word 'travelled' suggesting that 'sound travelling' might have been a phrase which children had encountered in a formal learning situation.

c. *Sound travel in the absence of media*

A substantial number of lower junior children who did specify a pathway for the sound suggested that it could only travel when there was nothing in its way to impede it. For example, rather than travelling through a medium such as the air or the string on a string telephone, the sound went through cracks around doors and windows, or through holes in the yogurt cartons of the telephone. Thus the sound had to be without a medium in order to be able to travel. The example in Fig. 3.37 shows how a child considered the message to go through the string telephone via a hole in the centre of the string. The notion that the sound needed to travel along an unimpeded path was also mentioned in Fig. 3.38 in connection with the drum. The sound was thought to leave the drum through a hole in the side of the drum.

Fig. 3.37

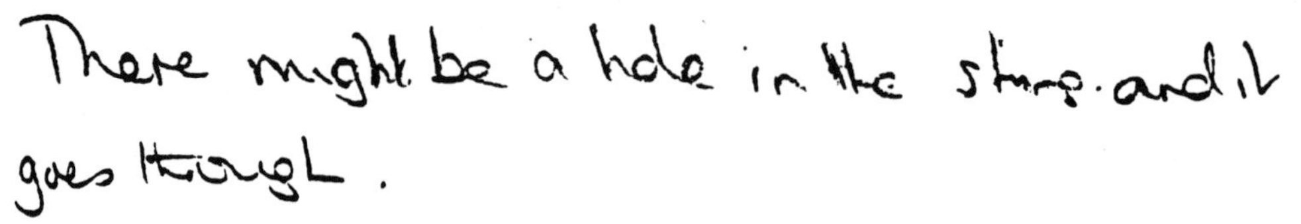

There might be a hole in the string. and it
goes through .

(Age 7 years)

"There might be a hole in the string and it goes through." (Teacher's annotation)

Fig. 3.38

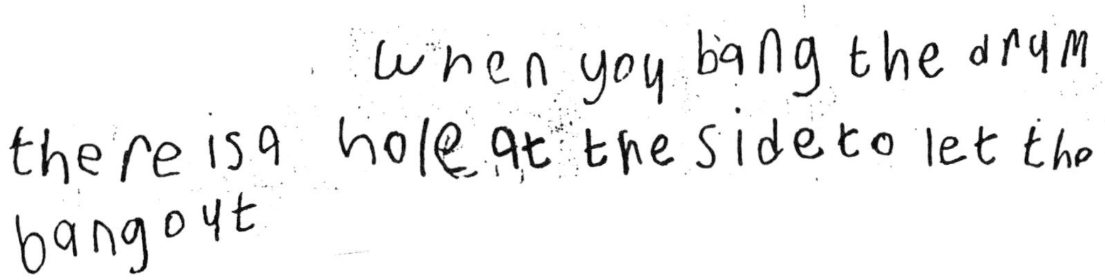

when you bang the drym there is a hole at the side to let the bang out

(Age 6 years)

"When you bang the drum there is a hole at the side to let the bang out."

Fig. 3.39

When the string is pulled tight the sound Kelly makes sort of jumps along the string.
Sound jumps along string goes into cup and then into your ear

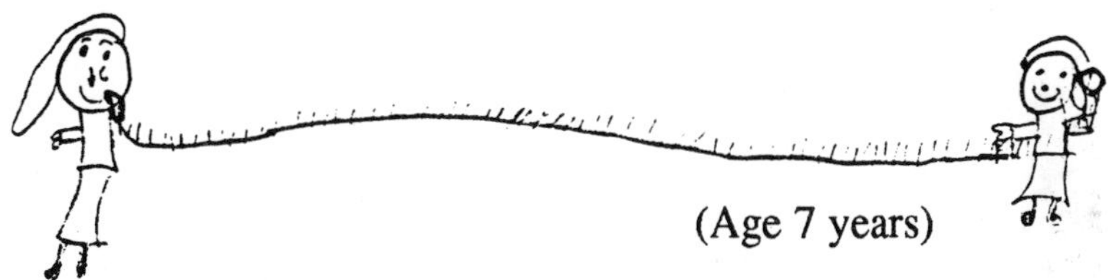

(Age 7 years)

"When the string it pulled tight the sound Kelly makes sort of jumps along the string. Sound jumps along string goes into cup and then into your ear." (Teacher annotation)

The notion of the need for an unimpeded pathway for sound travel might be intuitively linked with other experiences of travelling, since movement would be impeded by anything in the path of a moving object. It is possible that sound was envisaged as an invisible object with dimensions whose passage from source to receiver needed space. Children's notions of the nature of air also have a bearing here, since it would be possible for 'air' simply to be a label for the empty space around rather than implying a colourless, odourless substance with mass, volume and density. Were this the case, mention of sound travelling through air would also be an example of sound moving through an empty space rather than through a medium.

Figs. 3.37 and 3.39 illustrate two differing views about the role of the string in the string telephone. The advantages to these children of having the string are not clear but it does not appear to be perceived as a medium for transmission.

d. Sound travel through string

Many children did seem to consider the presence of a string, the explicit link in the string telephone between sound source and receiver, to explain the sound travelling. The number of children who expressed the idea of a medium through which sound travelled was greater with this activity than any other, with one third of the lower juniors (n = 34) and a half of the upper juniors (n = 40) responding in this way.

A detailed inspection of the children's drawings of themselves using the string telephone revealed large differences between upper and lower juniors. Over half of the lower juniors drew pictures in which their portrayal of the string telephone suggested an arrangement of the apparatus which would not carry a message from one person to another (Figs. 3.40, 3.41).

Fig. 3.40

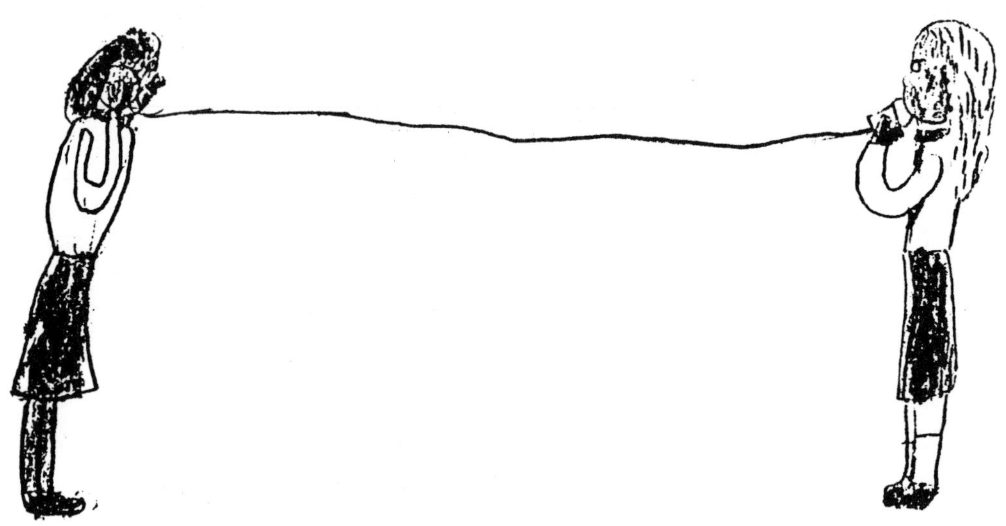

(Age 7 years)

Fig. 3.41

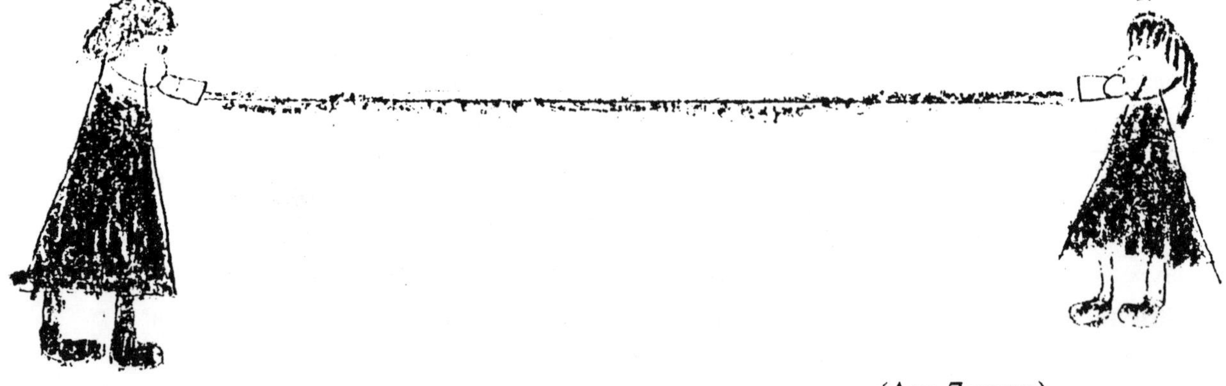

(Age 7 years)

This age-related difference in the accuracy of representation is ambiguous and could imply that the younger children's drawing skills were less developed or that the importance of accuracy was not perceived. Alternatively, there could be a difference in the perceptions of the role of the string between upper and lower juniors. The portrayal of two children facing each other (Fig. 3.40), of communicating through a slack string and of sending and receiving messages with both cups to the mouth (Fig. 3.41) and other responses which showed inaccurate representations of the string telephone and the two communicators were common only in the lower junior age range. This could suggest that children of that age have had sufficient experiences of sound to suggest that the string is responsible for carrying the message, but that they might not have been aware of which properties of the string were necessary conditions for it to transmit sound. Upper juniors expressed ideas which clarified these necessary properties by stating when the string telephone system would not work and thus also when it would (Fig. 3.42).

Fig. 3.42

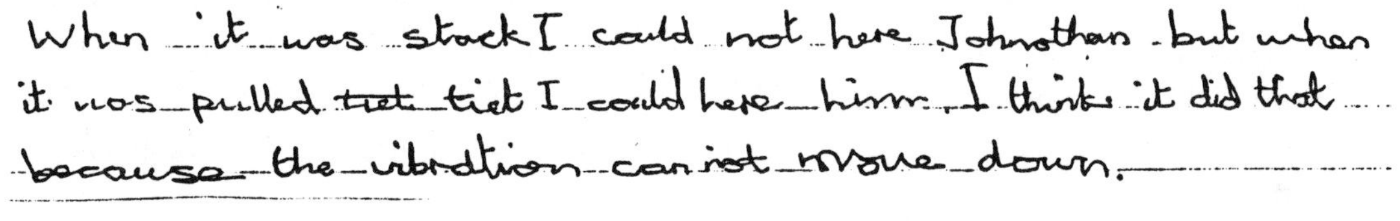

When it was stack I could not here Johnathan but when it wos pulled tiet I could here him. I think it did that because the vibration can not move down.

(Age 10 years)

"When it was slack I could not hear Jonathan but when it was pulled tight I could hear him. I think it did that because the vibration cannot move down."

e. *Sound travel through air*

The other Exploration activities elicited far fewer mentions of sound travelling through a medium, and the predominant view, particularly amongst lower juniors, was that sound needed to travel without a medium so that its passage was unimpeded. The air was mentioned by a small number of children but it was, in many cases, unclear whether the air was considered to be a medium or a space. Some children have attempted to portray air and sound entering the 'funnel' ear trumpet (Fig. 3.43), thus clarifying their intentions.

Fig. 3.43

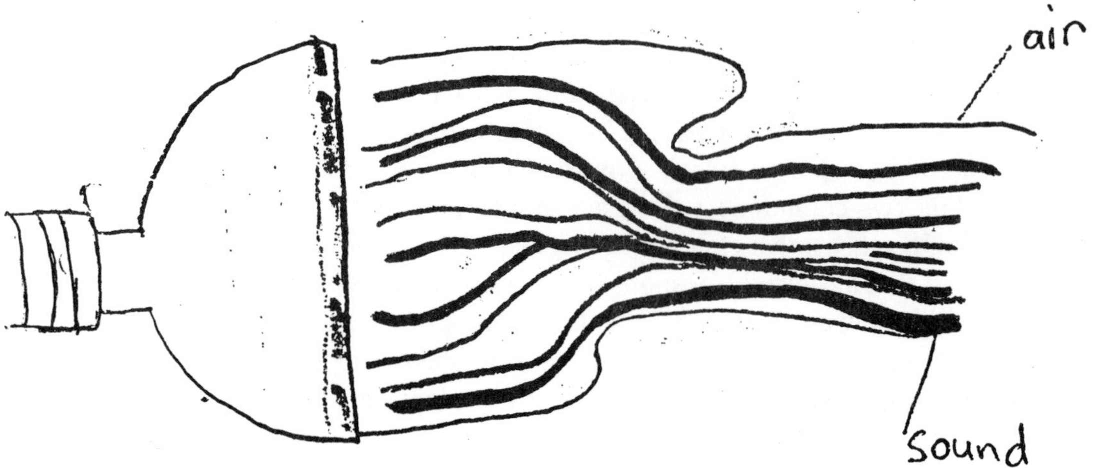

(Age 7 years)

Whether the transmission of sound through the medium of air is intuitively accessible to children, or whether it is an idea which has been acquired from secondary sources is not clear. The example in Fig. 3.44 is suggestive of the 'bell jar and alarm clock' demonstration in which the bell jar is gradually evacuated, making the alarm clock inaudible.

Fig. 3.44

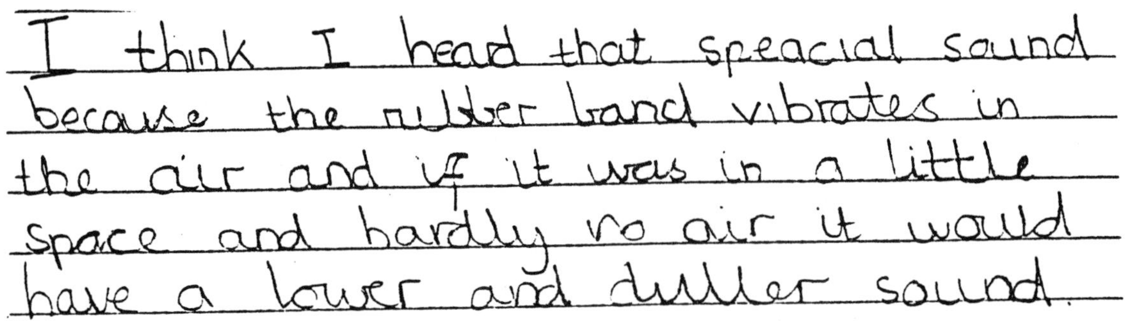

I think I heard that speacial sound because the rubber band vibrates in the air and if it was in a little space and hardly no air it would have a lower and duller sound.

(Age 9 years)

"I think I heard that special sound because the rubber band vibrates in the air and if it was in a little space and hardly no air it would have a lower and duller sound."

3.1.4 The Reception of Sound

Children's ideas about the reception of sound provided information about two distinct processes: the manner in which the ear received sound, and the process by which sounds were perceived once they had entered the ear. These two processes will be discussed in turn.

a. Sound reception by the ear

Children of junior age were more likely than infants to refer to their ears as the part of their body which was involved with receiving sound. Figs. 3.45 and 3.46 are examples of responses from younger children who either indicated no understanding of sound reception or who explained it without mentioning the ear.

Fig. 3.45

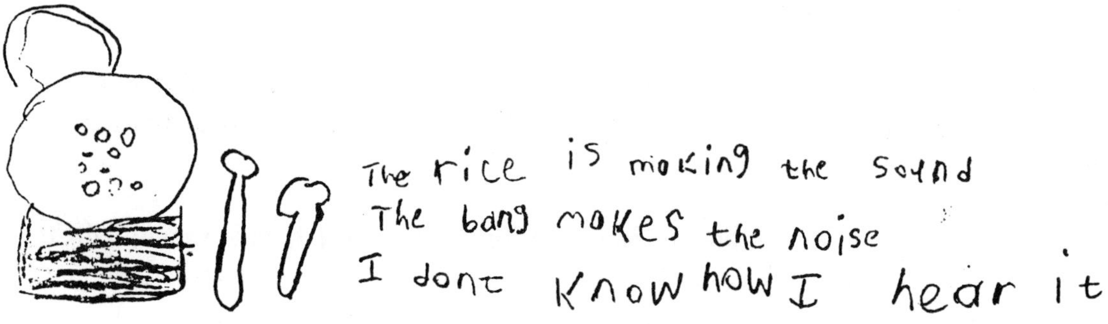

(Age 6 years)

"The rice is making the sound. The band makes the noise. I don't know how I hear it."

Fig. 3.46

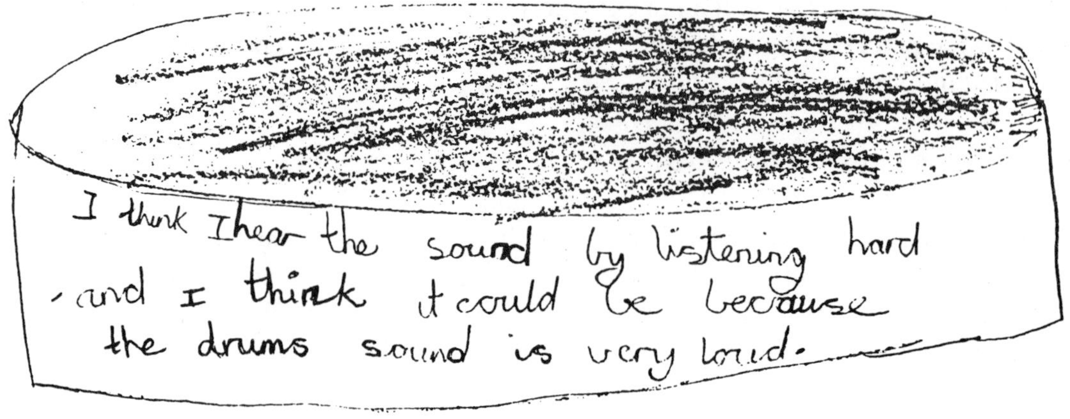

(Age 7 years)

"I think I hear the sound by listening hard and I think it could be because the drum's sound is very loud."

The Exploration activity involving the funnel ear trumpet encouraged children to express their ideas about how the ear trumpet affected hearing and it also provided some insights into the perceived nature of sound.

A prevalent response from infant and lower junior children was that the funnel helped them to hear better, but with no further explanation. Fig. 3.47 draws an interesting parallel with the ear trumpet activity from the child's own experiences.

Fig. 3.47

(Age 7 years)

"I can hear better with a funnel. Deaf people can. You just put it to their ear and say something through the other hole."

This unelaborated response of the funnel simply enhancing hearing was rarely given by upper juniors who tended to suggest more detailed mechanisms which could have created the auditory effects which they had noticed. The perception of distorted sounds appeared to lead some children to suggest mechanisms for sound collection which would explain the distortion rather than the apparent amplification which was expected. However, three detailed mechanisms were suggested, mainly by upper junior children. These were:

i. The funnel as a collector
ii. the funnel as a sound box
iii. The effects of pressure and compression of sound

i. The funnel as a collector

The simplest notion which was mentioned by some children of every age was that the ear trumpet enabled more sound to enter the ear because the opening of the funnel was wider than the ear. This type of explanation relates directly to the properties of the funnel which would be most obvious were a funnel observed in everyday use: the funnel is used to guide a large volume of material, for example, water, through a narrow opening like the neck of a bottle in the same way that sound is thought to be guided into the ear.

Fig. 3.48

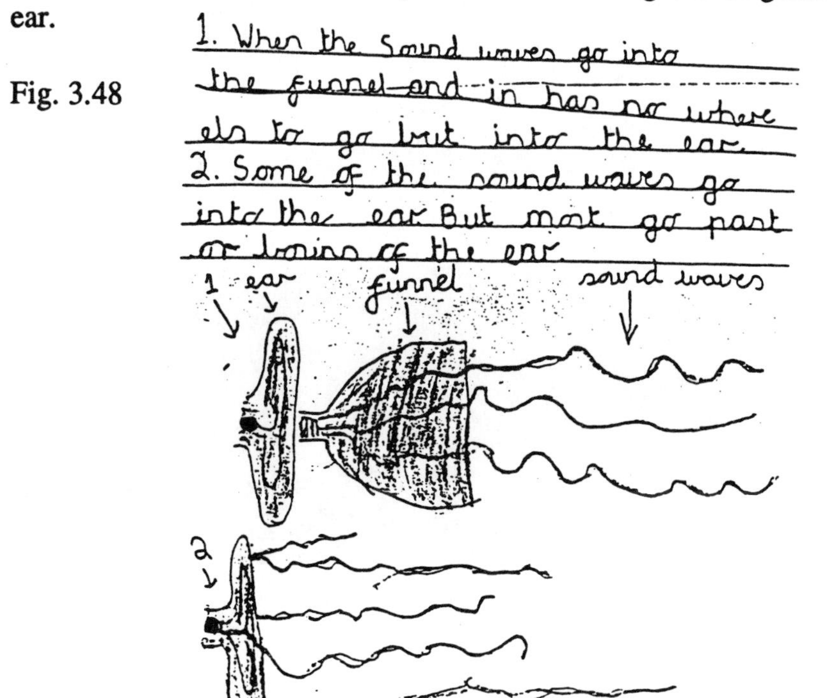

(Age 8 years)

"1. When the sound waves go into the funnel and it has nowhere else to go but into the ear.
2. Some of the sound waves go into the ear. But most go past or bounce off the ear."

Fig. 3.48 illustrates how one child did not seem to see the external ear as a smaller funnel, also capable of collecting sound; the drawing suggests that if the sound did not line up exactly with the hole into the ear, then, rather than being collected by the external ear, it was bounced back and lost. It is possible that the most perceptually obvious feature of the funnel was the wide opening, while the ear was considered to be flat rather than a very shallow dish which itself had a large mouth.

ii. The funnel as a sound box

The second mechanism which children suggested was that the funnel provided surfaces off which the sound could bounce. This description is very close to the conventional explanation of one of the ways in which the funnel did affect the manner in which sounds were heard.

Fig. 3.49

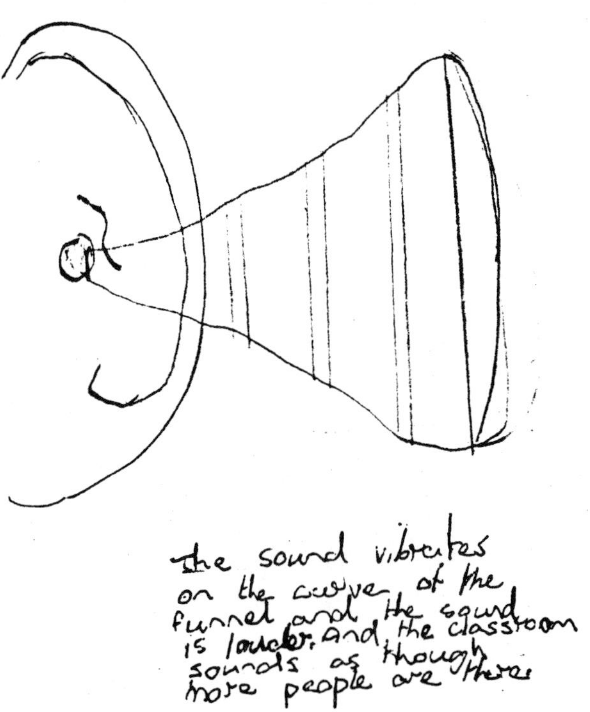

(Age 8 years)

"The sound vibrates on the curve of the funnel and the sound is louder, and the classroom sounds as though more people are there."

Fig. 3.50

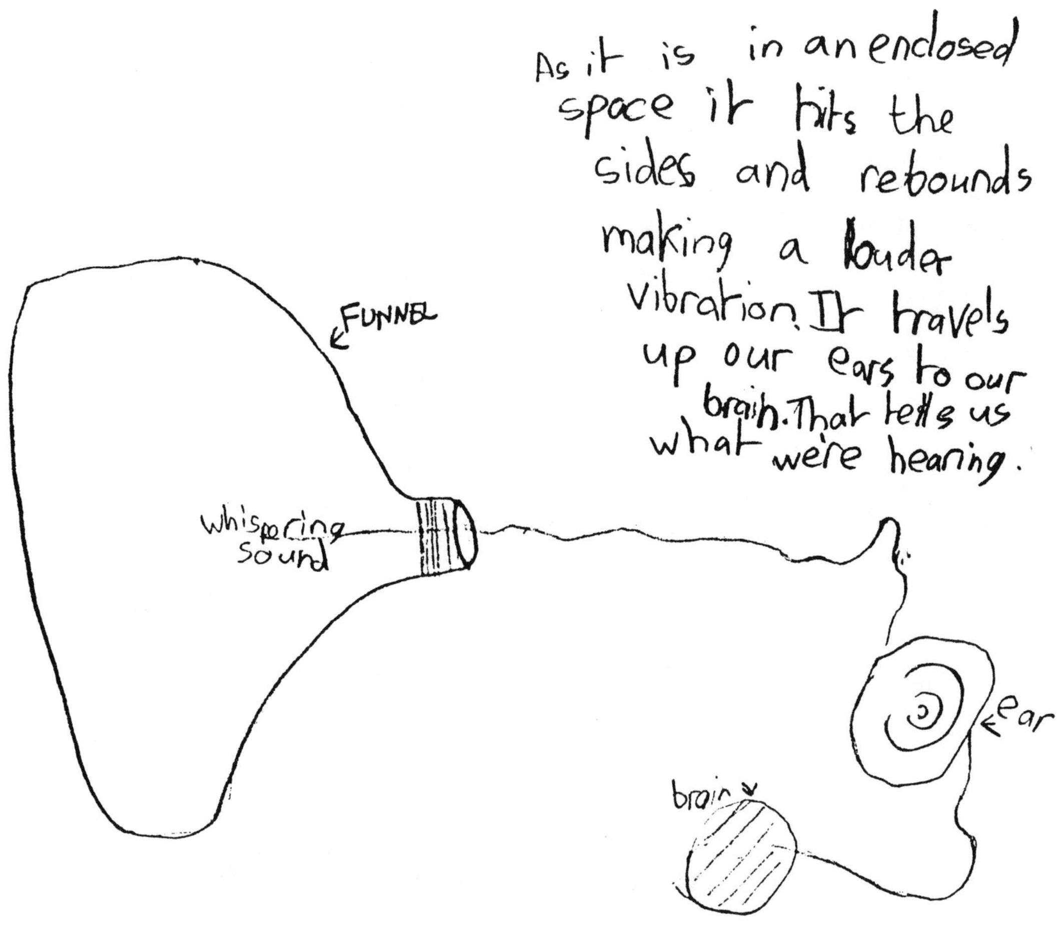

(Age 10 years)

"As it is in an enclosed space it hits the sides and rebounds making a louder vibration. It travels up our ears to our brain. That tells us what we're hearing."

It is possible that the example shown in Fig. 3.50 was suggesting that sound multiplied each time it rebounded, and the existence of more copies of the same sound would mean that the cumulative result would be a louder sound. Those children who suggested sound being less clear might have felt that the existence of more copies of the sound would lead to them all being jumbled up rather than them adding neatly together to enhance the sound. A correctly proportioned ear trumpet should not have produced the effect of muffling the sound.

iii. The effects of pressure and compression of sound

The third mechanism was suggested as another explanation for less clearly heard sounds. Approximately one-tenth of upper juniors thought that the narrowing of the funnel would result in the sound being distorted to fit through the opening into the ear.

Fig. 3.51

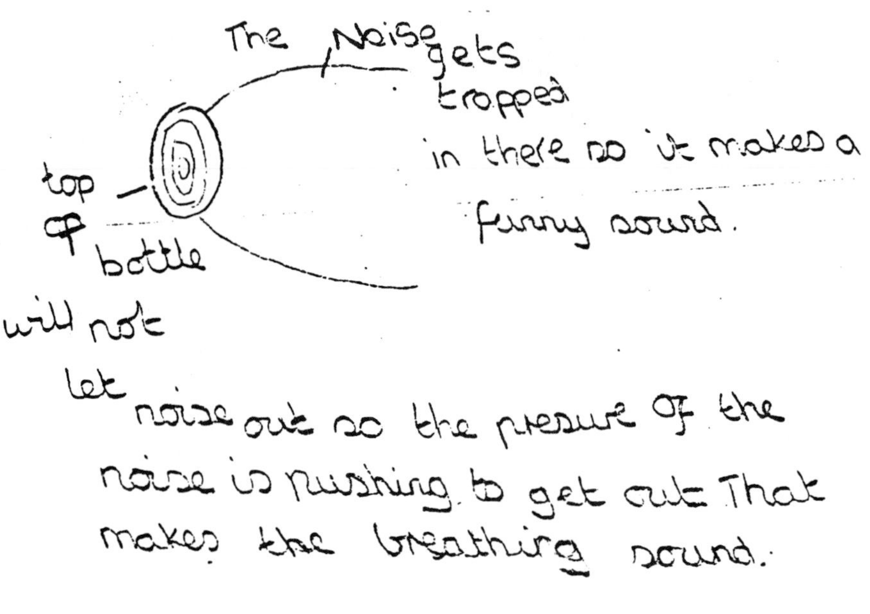

(Age 9 years)

"The noise gets trapped in there so it makes a funny sound. Top of bottle will not let noise out so the pressure of the noise is pushing to get out. That makes the breathing sound."

Fig. 3.52

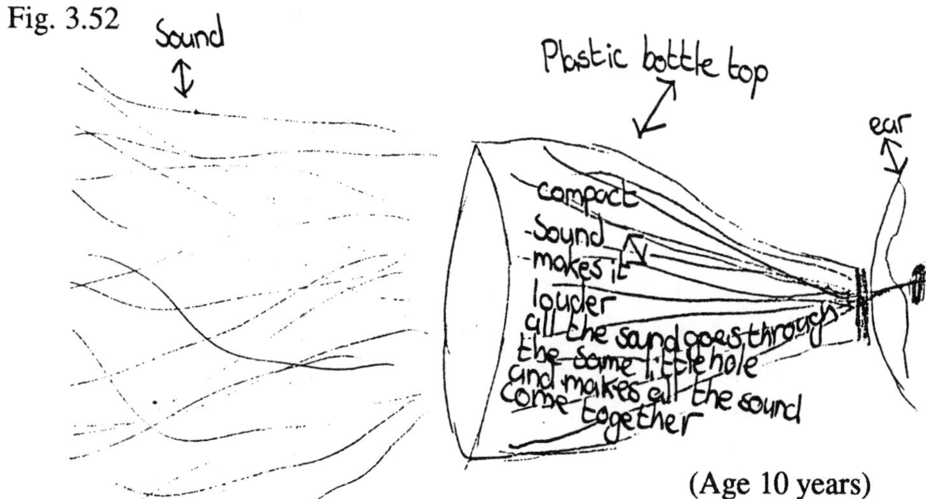

(Age 10 years)

"Compact sound makes it louder. All the sound goes through the same little hole and makes all the sound come together."

These examples in which children mention pressure or compression are fairly close to the scientific description of the way in which sound is amplified by the funnel. Whereas in Fig. 3.51 the reasoning is in terms of distortion of the sound, Fig. 3.52 is an excellent description of the phenomenon phrased in intuitive language.

b. *Sound perception after sound has entered the ear*

The idea that sounds are not heard simply by the external ear is a complex one and is dependent to a large extent on learning from secondary sources. As previously mentioned, only one-third of infant children mentioned ears in connection with hearing. A few responded in a manner similar to Fig. 3.53, that they hear with their ears.

Fig. 3.53

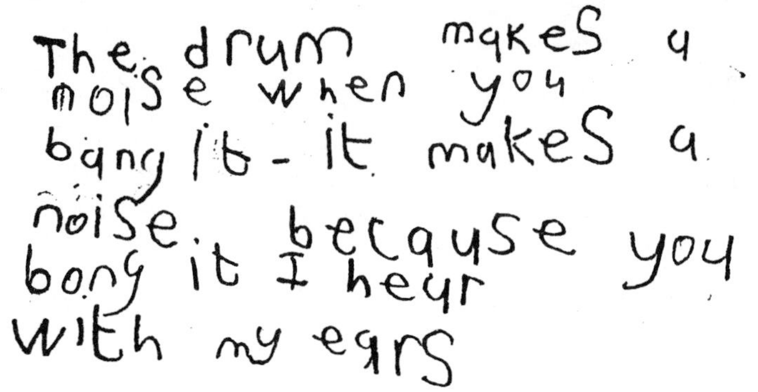

The drum makes a
noise when you
bang it - it makes a
noise because you
bong it I hear
with my ears

(Age 7 years)

"The drum makes a noise when you bang it - it makes a noise. Because you bang it I hear with my ears."

The apparatus contained within the inner ear to translate sound vibrations into neural impulses which travel to the brain was not mentioned in detail by any child. The ear drum was mentioned by one-fifth of upper juniors (n = 84) and one-tenth of lower juniors (n = 74), and one child described bones within the ear (Fig. 3.54).

Fig. 3.54

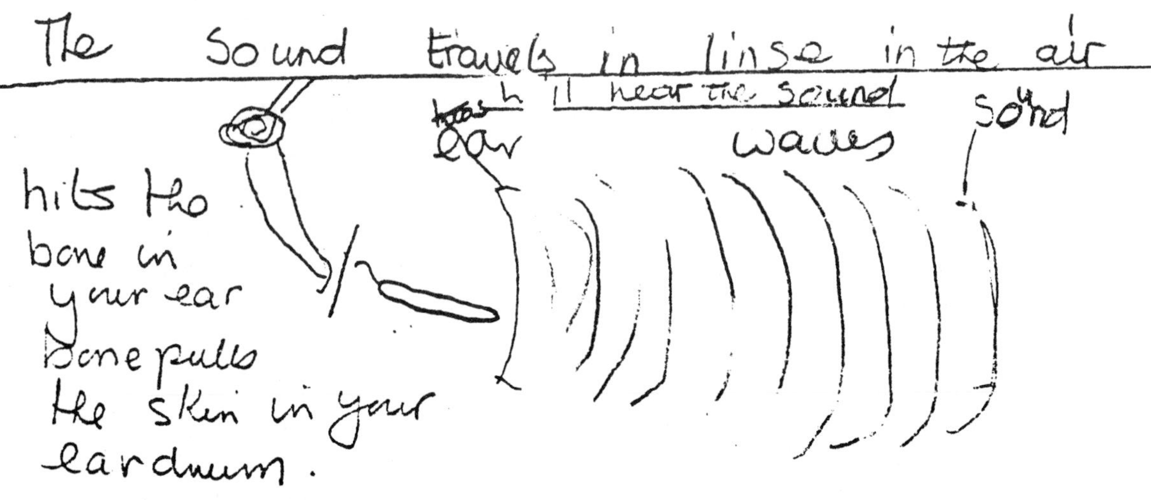

(Age 10 years)

"The sound travels in lines in the air - hits the bone in your ear-bone pulls the skin in your eardrum." *(Some teacher annotation)*

The role of the ear drum and its appearance were speculated upon by some children (Figs. 3.55 and 3.56) and the brain was mentioned by one-tenth each of lower and upper juniors. The ear drums in Fig. 3.55 appear to be very small versions of the musical instrument with which the word 'drum' would be associated.

Fig. 3.55

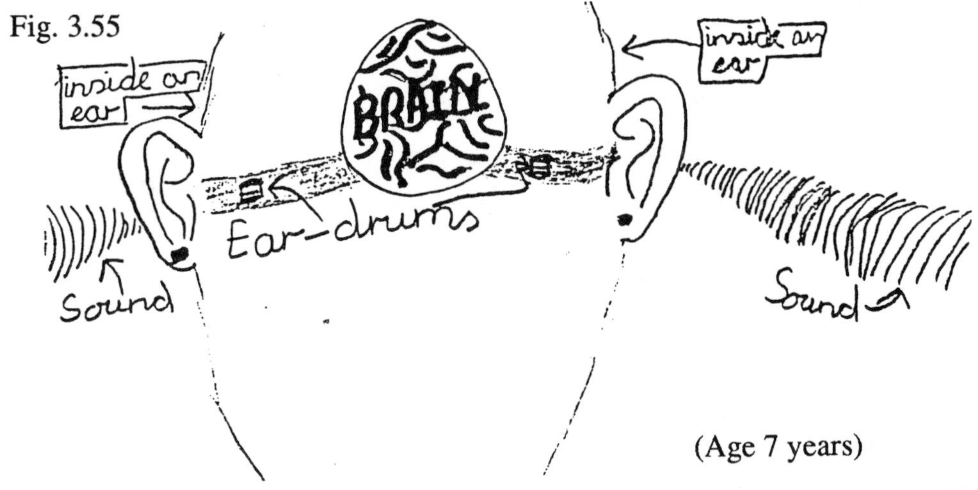

(Age 7 years)

Fig. 3.56

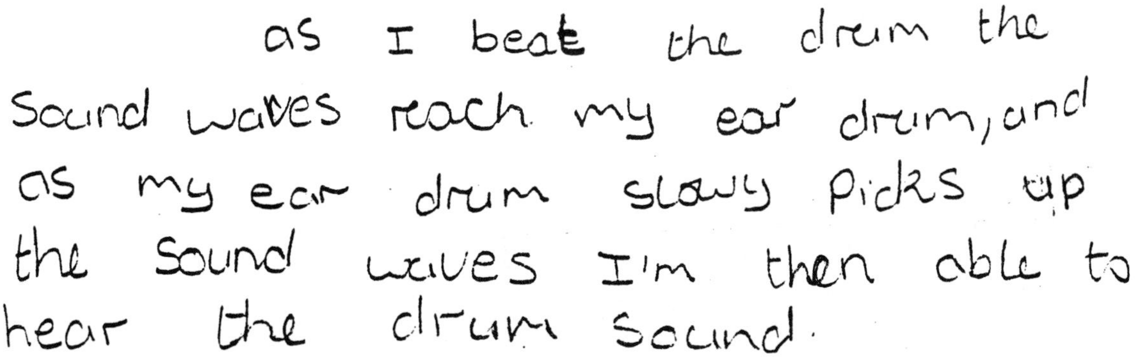

as I beat the drum the sound waves reach my ear drum, and as my ear drum slowy picks up the sound waves I'm then able to hear the drum sound.

(Age 10 years)

"As I beat the drum the sound waves reach my ear drum, and as my ear drum slowly picks up the sound waves I'm then able to hear the drum sound."

The need for sound to enter the ear was mentioned by only one infant. A substantial number of junior children considered the form taken by sound when it reached the ear drum and the brain, and some quite elaborate pathways were suggested. In the junior age range, one-tenth of lower juniors and half of the upper juniors reported sound entering the ear. The upper junior sample was divided between those children describing 'sound' and those who talked about 'sound waves' or 'vibrations'. Over a quarter of the upper juniors suggested vibrations. The notion of vibrations being set up in the ear drum (including the sound banging on the ear drum) was also given only by upper juniors. Vibrations/sound waves were mentioned by approximately 11% of the whole sample (n = 184) in connection with hearing sound.

Fig. 3.57

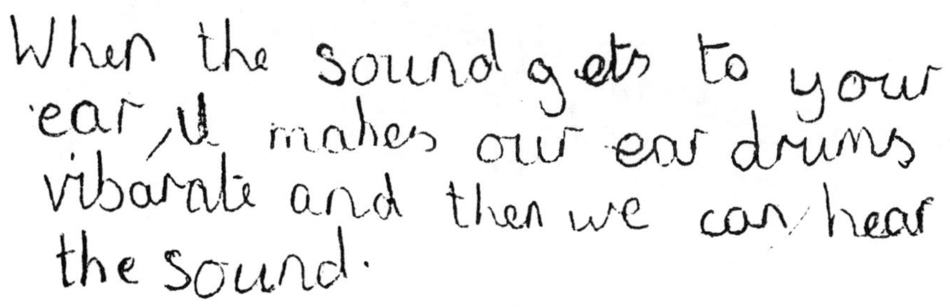

When the sound gets to your ear, it makes our ear drums vibarate and then we can hear the sound.

(Age 10 years)

"When the sound gets to your ear, it makes our ear drums vibrate and then we can hear the sound."

Fewer children included the step from ear drum to brain in the mechanism and the majority of those who did were upper juniors. Equal numbers (approximately 5%) of these juniors considered that the signal went to the brain in the form of 'sound', vibrations (Fig. 3.58) or some other form of signal (Fig. 3.59).

Fig. 3.58

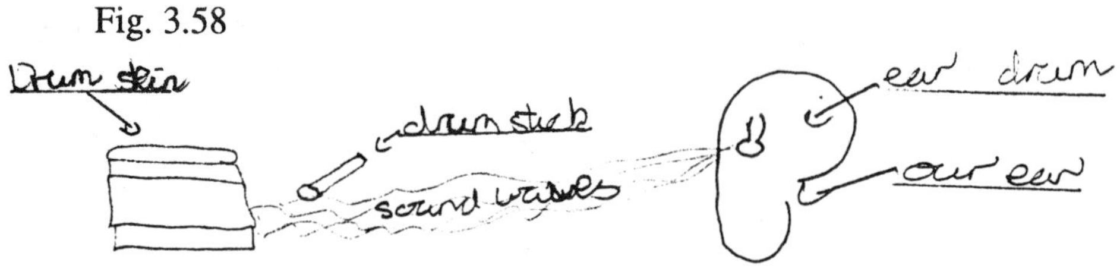

When the air goes into our ear and hits our ear drum. The ear drum vibrates all the way to the brain then the brain translates it.

(Age 10 years)

"When the air goes into our ear and hits our ear drum, the ear drum vibrates all the way to the brain then the brain translates it."

Fig. 3.59

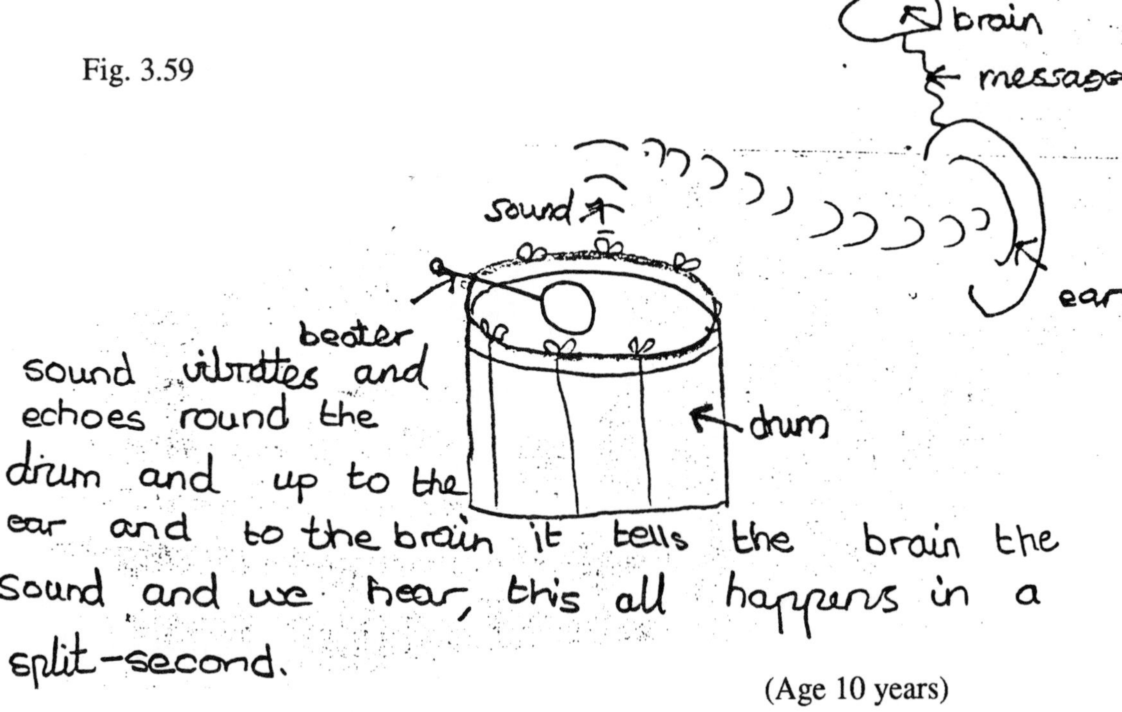

"Sound vibrates and echoes round the drum and up to the ear and to the brain. It tells the brain the sound and we hear, this all happens in a split-second."

Fig. 3.58 also indicated that the brain was thought to undertake a translation of the signal it received, and this final step was suggested by a small number of upper juniors.

3.1.5 Children's Representations of Sound

The appearance on diagrams of any representations of sound was more common the older the children and also appeared to be more common for boys than girls. In the upper juniors over half of the sample (n = 66) used a representation, three quarters of whom were boys. The four main issues which the diagrams informed were:

a. Whether the sound was portrayed as continuous or discontinuous from source to receiver.

b. Whether the sound was depicted as lines approximately parallel or perpendicular to the shortest line from sound source to receiver.

c. Whether the sound diverged as it travelled or remained on a path of constant width.

d. The form of notation chosen by the child to represent the sound.

a. Continuous or discontinuous sound

Children either showed the sound as happening once, as a discrete entity which then travelled to the receiver (Fig. 3.60) or as a continuous line from source to receiver (Figs. 3.61, 3.62).

Fig. 3.60

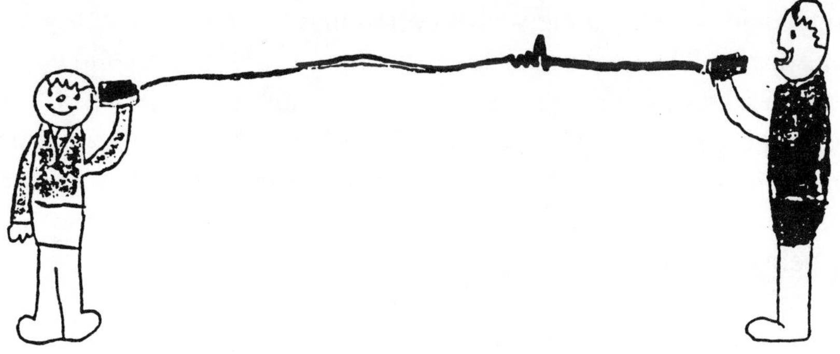

(Age 7 years)

Fig. 3.61

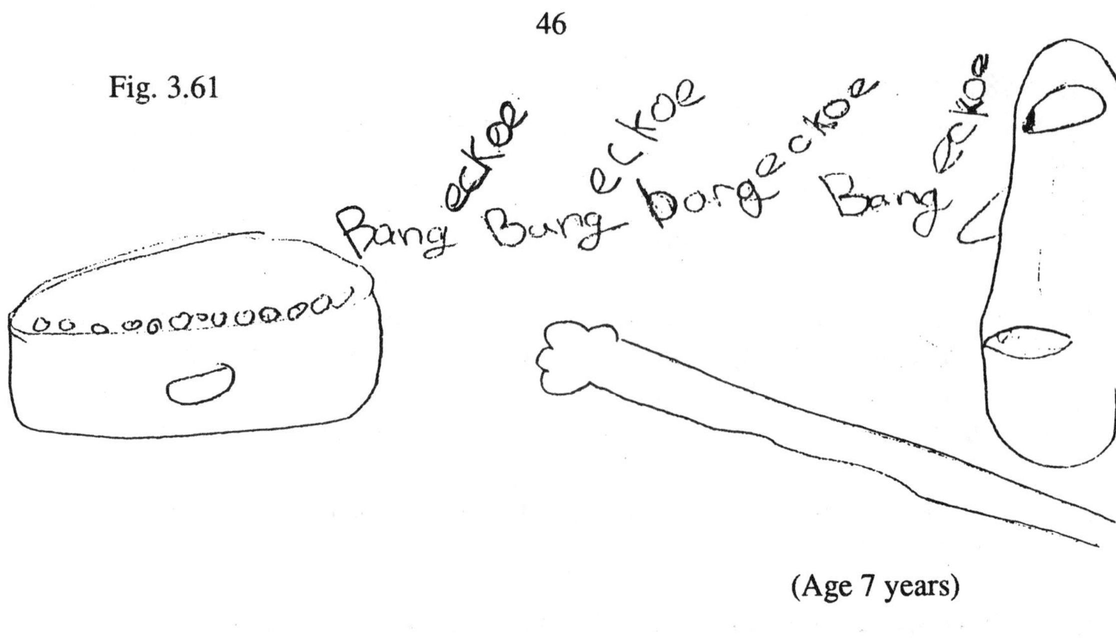

(Age 7 years)

Fig. 3.62

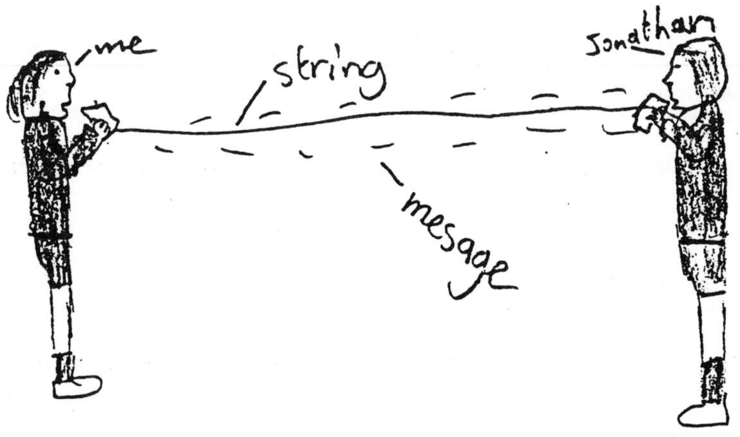

(Age 9 years)

The discrete representation of sound could be considered to be an intuitively governed response to the question of how to show sound travelling. This form of representation would be compatible with a notion that 'sound' is an entity which travels through space. The continuous representation is ambiguous and could be interpreted in several ways. The idea that sound is a continuous series of vibrations would not be very accessible to children since sound travelling can neither be seen nor easily timed. While it is possible that the children were portraying a series of vibrations it seems likely that children were attempting to portray the position of the sound over a period of time.

b. *Parallel or perpendicular orientation of sound lines*

The most common depiction of sound for both lower and upper juniors was in a direction broadly parallel to the direction of sound travel (Fig. 3.63). This was the only type of representation elicited from lower juniors.

Fig. 3.63

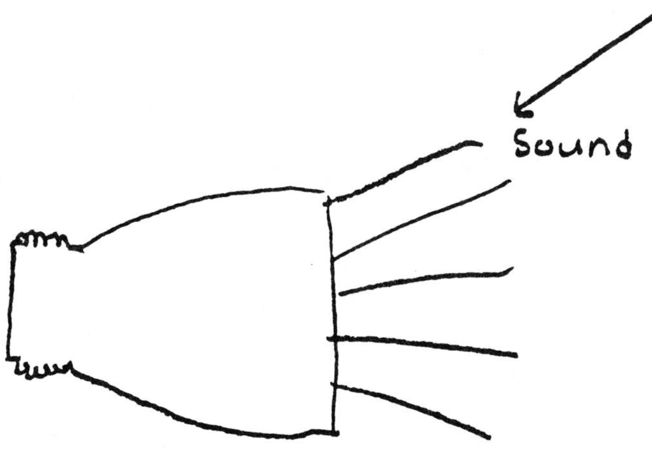

(Age 9 years)

A quarter of the upper juniors who used a representation showed sound as lines perpendicular to the direction of sound travel (Fig. 3.64). This suggests a response which has been learned from secondary sources since the intuitive response would seem more likely to indicate the direction of travel rather than the wave form.

Fig. 3.64

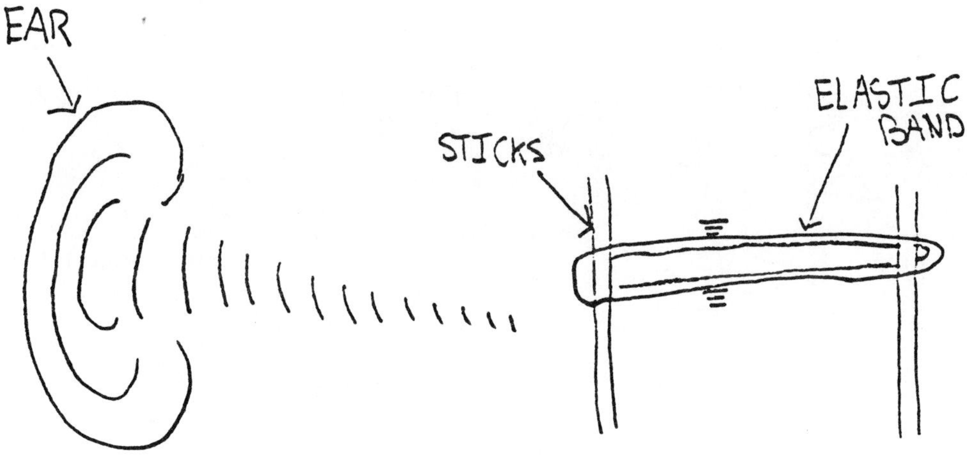

(Age 10 years)

c. Divergent or non-divergent paths of sound travel

The notion that sound spreads out from its source was held by some children (Fig. 3.66) and was found in conjunction with sound depicted both parallel and perpendicular to the direction of sound travel. A large number thought that the sound went only to the intended listener, and remained in a path of constant width (Figs. 3.66 and 3.67). This latter response may be a component of the active listening model whereby the sound goes straight to the person who is attending to it; it might be regarded as an egocentric type of response.

Fig. 3.65

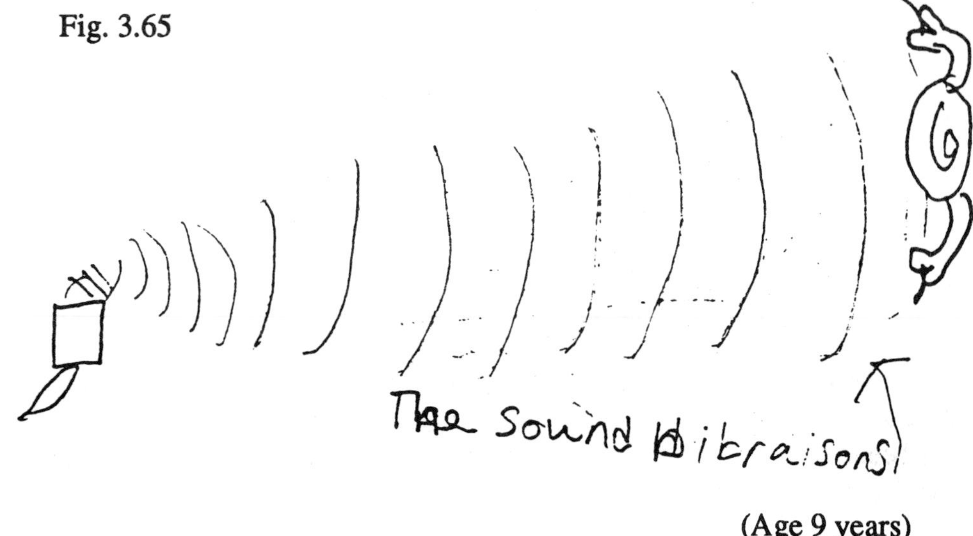

(Age 9 years)

Fig. 3.66

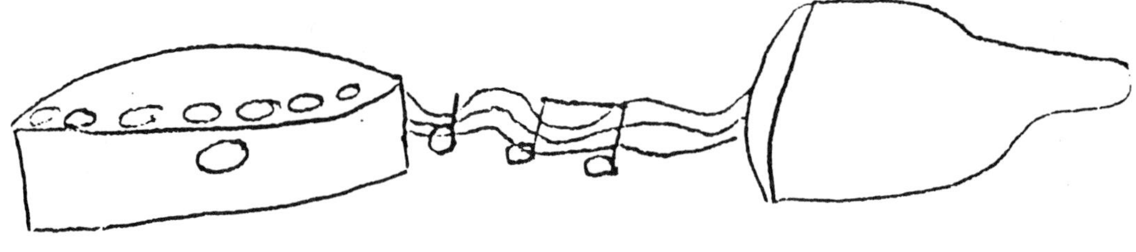

(Age 10 years)

Fig. 3.67

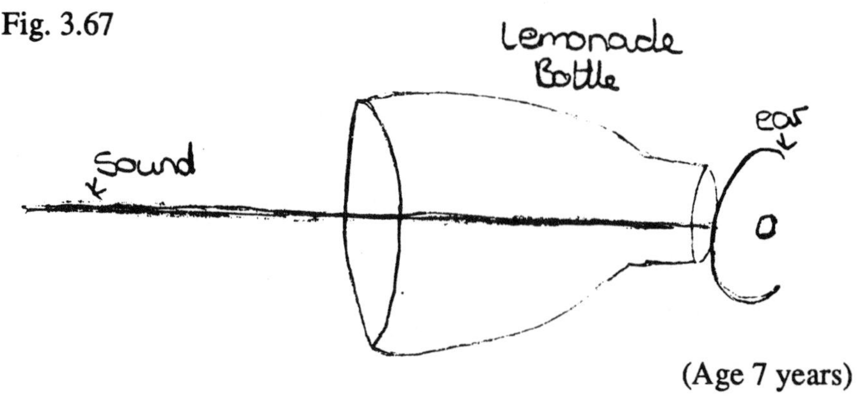

(Age 7 years)

The illustrations shown in Figs. 3.66 and 3.67 were each drawn by children who suggested that the ear trumpet affected the way they heard, but their representations of sound do not make any effect clear: the sound appears to pass unimpeded through the centre of the funnel and into the ear.

One child depicted sound converging from the source, a drum, which might again suggest that some secondary source has led the child to make a response which had not been completely assimilated (Fig. 3.68).

Fig. 3.68

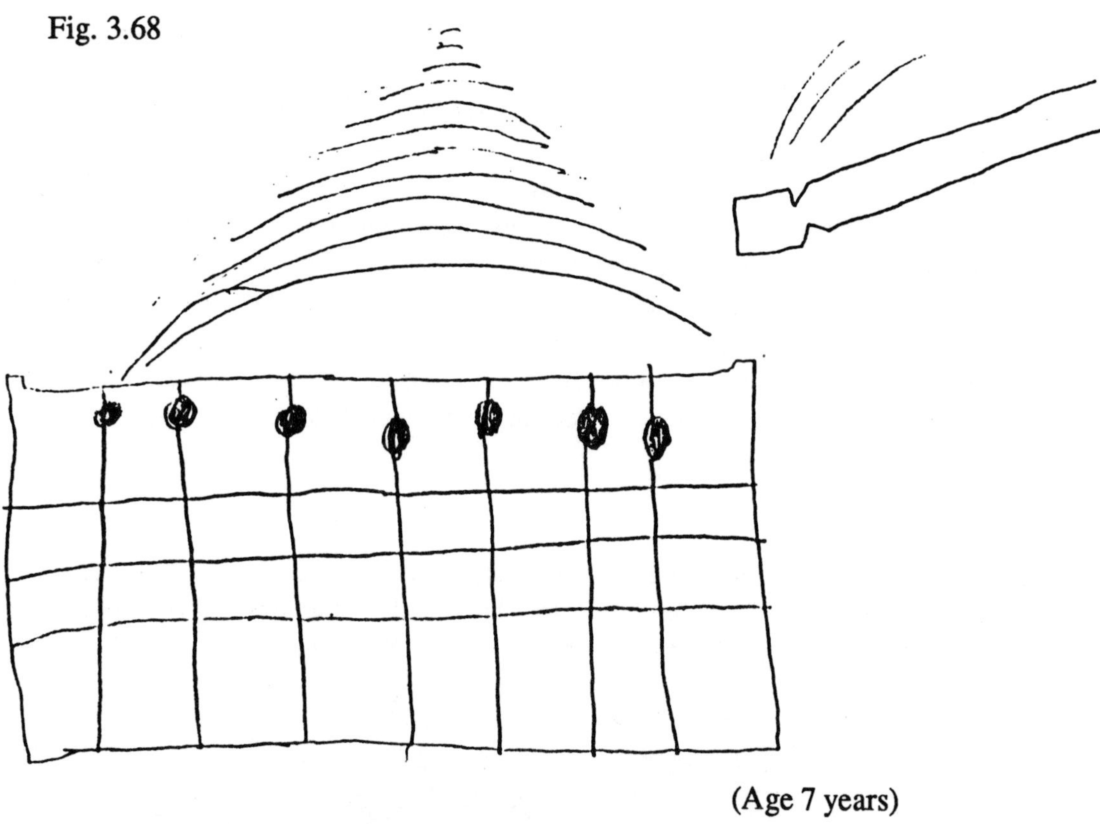

(Age 7 years)

d. Notation used for sound

A wide range of notations was used to represent sound. Words were used (Fig. 3.69), as well as lines (Fig. 3.68), arrows, musical notes (Fig. 3.70), and shading.

Fig. 3.69

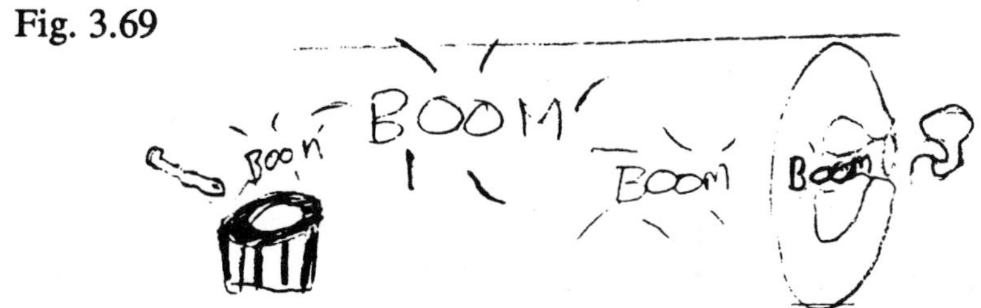

(Age 10 years)

Fig. 3.70

(Age 10 years)

The interpretation of these different notations of sound is open to speculation. The formal representations, such as perpendicular lines, seem likely to be used in what the child remembers to be the accepted manner from pictures or diagrams which have been seen. The more intuitive notations, the words and musical notes could also have been acquired by children from other, less formal secondary sources such as comics or they could demonstrate the child's own attempts to represent sound. An inspection of some children's comics revealed examples of all of the notations used by children, except for sound waves, perpendicular to the direction of sound travel. In comics, these perpendicular lines appear to denote movement rather than sound.

3.1.6 Vocabulary

The word 'vibration' was used by a very large number of children and it appeared to be used to mean several different things. The child whose drawing is shown in Fig. 3.71 appears to use 'vibrate' to mean 'does not resonate'.

Fig. 3.71 I stretched a rubber band
between my finger and
thumb and plucked it with
my other hand. I noticed
- it dosn't make very much
noise, it vibrats and it hurts
my hand. I can't play a
tune. I plucked a rubber
band that was on a box
I noticed - it makes a
loud noise and it dosn't
vibrat. I can play a tune

(Age 7 years)

"I stretched a rubber band between my finger and thumb and plucked it with my other hand. I noticed - it doesn't make very much noise, it vibrates and it hurts my hand. I plucked a rubber band that was on a box. I noticed - it makes a loud noise and it doesn't vibrate. I can play a tune."

Some children appear to use 'vibrating' as an adjective (Fig. 3.72) or a verb (Fig. 3.73) to qualify the word 'sound'. This usage of 'vibrating' as a qualifying term suggests that the children's definition of 'vibration' is restricted to particular types of sound, possibly where the movements of the sound producer are visible, as with the rubber band.

Fig. 3.72

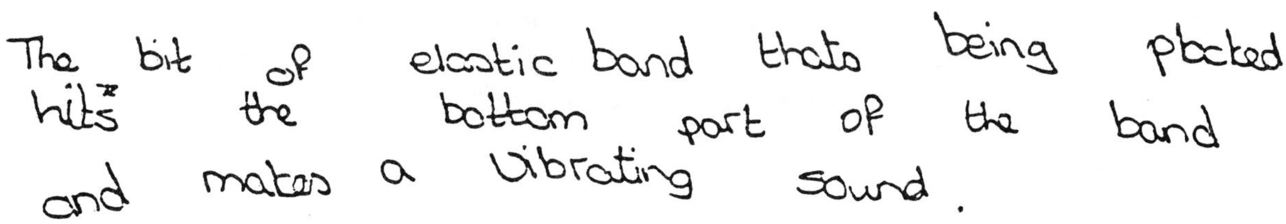

The bit of elastic band thats being plcked hits the bottom part of the band and makes a vibrating sound.

(Age 10 years)

"The bit of elastic band that's being plucked hits the bottom part of the band and makes a vibrating sound."

Fig. 3.73

By the elstic band moving it makes a noise by vibrating the sound, If tighter band you get a high note If you slacken the band you will get a lower note.

(Age 9 years)

"By the elastic band moving it makes a noise by vibrating the sound. If tighter band you get a high note. If you slacken the band you will get a lower note."

'Vibration' also seems to be used to imply repetition so that children attempting to understand the notion of sound travelling might be doing so by having the sound repeat its way along.

The word 'echo' was used by upper and lower junior children in connection with several of the Exploration activities, particularly those which incorporated a resonating chamber - the funnel and the drum and the rubber band on a box. A commonly expressed definition of 'echoing' was that the sound was repeating itself (Fig. 3.61, p. 34.) Perceptually, an echo would intuitively suggest a repetition of the initial sound. Some understanding of sound transmission would be necessary for children to consider sound to bounce back from a surface.

3.1.7 Summary

1. 'Sound' and 'vibration' were not intuitively linked by young children but an association developed as children's experiences broadened. Whether sound caused vibrations or vibrations caused sound seemed to depend upon the context. Some children suggested that sound and vibration were the same.

2. The production of sound from an object was often attributed to the properties of the object or to an impact. Children suggested mechanisms for the generation of sound from a drum. These mechanisms often involved vibrations and the site of sound production was either inside the drum or at the surface.

3. Sound transmission was not an idea which was expressed by many young children. Infant children said they heard sounds because of the volume, the proximity to the sound source or because of a characteristic of the listener. Where sound travel was mentioned there was a prevalent idea that sound needed an unobstructed path along which to travel. Some older children considered sound to travel through the string on the string telephone, or through air. What was meant by 'air' was often unclear.

4. Sound reception was frequently associated with the ear. A funnel ear-trumpet was thought to affect hearing by either collecting orconcentrating sound so that more reached the ear. A small number of junior children mentioned the ear drum and the brain in connection with hearing. Some of these children mentioned vibrations being set up in the ear drum.

5. A wide range of representations was used, some idiosyncratic and parallel to the direction of sound travel, and others more like the accepted scientific notation, perpendicular to the direction of travel. A small number of children portrayed sound spreading out from the source. The idiosyncratic notations could have been influenced by the representations of sound shown in children's comics.

6. Junior children made frequent use of words such as 'vibrate', 'echo', 'travel' and 'sound wave' in association with descriptions of ideas about sound. Both 'vibrate' and 'echo' were often used in a manner which implied a meaning of 'repeat'.

3: CHILDREN'S IDEAS

Part 2: Responses to Individual Interviews

A stratified sample of children from each of the classes involved in project work was selected to be interviewed individually both before and after the Intervention. This sample was balanced for gender, and was intended to contain approximately twenty-four children from each age band. However, since these interviews were long and time-consuming, constraints on the interviewers' time often meant that if a child was absent on the day set aside for interviewing, it was not possible to revisit the school on another occasion.

The pre-Intervention sample contained:-

16 infants (8 girls, 8 boys)
16 lower juniors (8 girls, 8 boys)
25 upper juniors (11 girls, 14 boys)
57 children (27 girls, 30 boys)

The pre-Intervention interviews were structured around the Exploration activities to which the children had access in their classrooms. These were:-

a. Sending and receiving a message through a string telephone
b. Stretching and plucking a rubber band
c. Listening to sounds through an ear trumpet
d. Hitting a drum which had some rice grains on the skin
e. Listening to everyday sounds

Each of these activities made some aspects of sound more explicit than others: the string telephone, for example, emphasised sound transmission. Because of these context-related effects, children were asked different questions for each activity and responses to all of these questions have been reported separately.

In order to be able to draw out some common threads between activities and to make comparisons, a small number of super-ordinate categories, common to all activities, have been developed. These categories are concerned with:-

sound production
sound transmission
sound reception

The structure of the report will be such that each activity will be discussed in turn, drawing upon data from the activity-specific categories. The activities will then be compared using the super-ordinate cagegories in order to compile a picture of children's ideas about sound.

a. Sending and receiving a message through a string telephone

The string telephone makes explicit a pathway for the transmission of sound, and this aspect of sound was emphasised in the elicitation. Children's responses have been scored to these questions:-

i. How does the message get to your partner?
ii. How does it help to have a string telephone?
iii. What does the string do (what qualities of the string are important)?

i. How does the message get to your partner?

The purpose of this question was to establish whether the children explained the sound being heard by their partner in terms of the sound travelling, or whether they used some alternative explanation. Children's responses fell into four main groups (though these categories were not mutually exclusive).

Table 4.1 Ideas about how a message reaches the listener when using a string telephone (Percentages)

	Infants (n = 16)	Lower Juniors (n = 16)	Upper Juniors (n = 25)
The listener was listening	19 (3)	6 (1)	- -
The string was pulled tight	25 (4)	-	-
The message went in and out through holes in the yogurt pots	25 (4)	19 (3)	4 (1)
The message went via the string	38 (6)	88 (14)	100 (25)
Don't know	6 (1)	6 (1)	-

The proportions of children in each of these groups varied with age: all of the upper junior children and the majority of lower juniors thought the message went via the string. The pattern of responses found among infant children was significantly different ($p < 0.001$): less than half of them considered the message to go via the string (38%). The remaining infant responses were relatively evenly distributed between each of the other categories, with equal numbers implicating the tightness of the string or replying that the sound went in and out through the holes in the yogurt pot. This latter response, also made by some juniors, could suggest that the children were developing an awareness that the message must be travelling. However, they appeared to feel that the message needed to travel through nothing, i.e. spaces or holes, rather than through something, i.e. the string. The remaining children considered the important factor to be that the child receiving the message was listening (infant 19%, lower junior 6%). This response suggests that the children are using a psychological model to explain being able to hear. This notion of 'active listening' implies the need for the listener to be attending to the message, rather than suggesting a pathway or mechanism by which the message could be heard.

b. *Stretching and plucking a rubber band*

The Exploration activity in which children were invited to engage with the rubber band by stretching it and plucking it emphases the production of sound but also provides an opportunity to ask children about sound transmission and reception. Children's responses have been scored to the following questions:

i. How do you think the rubber band makes a sound?
ii. How does the sound reach you so you can hear it?

i. How do you think the rubber band makes a sound?

This question probed the nature of sound production. As these categories are not mutually exclusive children may make more than one response.

Table 4.2 Ideas about how a rubber band makes a sound (Percentages)

	Infants (n = 16)	Lower Juniors (n = 14)	Upper Juniors (n = 22)
The band is rubber	31 (5)	43 (6)	5 (1)
The band needs to be plucked	69 (11)	79 (11)	86 (19)
The band needs to be stretched	19 (3)	14 (2)	32 (7)
The band hits the air and there's a sound	-	-	14 (3)
The band vibrates/wobbles etc.	6 (1)	36 (5)	41 (9)
Observation of the band moving (not necessarily connected with sound production)	25 (4)	50 (7)	59 (13)
Other	6 (1)	-	9 (2)
Don't know	6 (1)	-	-

Children in the two younger age groups quite commonly responded that the production of a sound was related to the band being rubber, in other words that the nature of the rubber band itself was responsible for a sound being made. This type of response was rarely given by upper junior children ($p < 0.01$) who were more likely to give responses relating to contingent conditions for the process of sound production. The need for the rubber band to be stretched was reported by one third of upper juniors and by smaller numbers of younger children and the importance of plucking the band was recognised by the majority of children interviewed from every age.

The number of children who observed the band moving doubled between the infants and lower juniors($p < 0.05$), and substantially more lower juniors than infant children mentioned that the band needed to vibrate (or some suitable synonym, such as wobble) for sound to be produced. When children were asked whether they would be able to tell that a rubber band was making a noise if they could not hear it, nearly all infant children, including three quarters of those who had observed the band moving, responded negatively. In both junior age groups the reverse happened, with nearly three quarters of children who observed the movements making a link between the vibrations and sound production. These findings indicate that there could be a hierarchy of observations which children need to make in order to develop a coherent generalizable concept of sound production. Until children have noticed the pertinent contingent variables they are unlikely to be able to suggest an answer which would not be specific to the context being explored.

ii. How do you think the sound reaches you so you can hear it?

This question probed the nature of sound transmission and reception in connection with the rubber band.

Table 4.3 Ideas about how sound from a plucked rubber band reaches the listener (Percentages)

	Infants (n = 14)	Lower Juniors (n = 14)	Upper Juniors (n = 21)
Sound is heard because of a characteristic of the listener	50 (7)	21 (3)	24 (5)
Sound reaches you (no further explanation)	- -	43 (6)	29 (6)
Sound travels via the air	- -	14 (2)	14 (3)
Other	- -	- -	5 (1)
Don't know	50 (7)	21 (3)	29 (6)

This table shows that there are age-related trends in response to this question. There seem to be large differences between the infants and juniors in the ideas concerning sound transmission ($p < 0.05$). Half of the infants who were interviewed did not know how the sound reached them, and the other half related hearing to a characteristic of themselves, their positioning or an action they had made.

Table 4.4 An itemization of different characteristics of the listener (Percentages)

	Infants (n = 14)	Lower Juniors (n = 14)	Upper Juniors (n = 21)
You listen carefully	14 (2)	-	-
You hear it	14 (2)	7 (1)	-
It's loud	-	-	5 (1)
You pull it a long way	-	-	5 (1)
You've made a noise	14 (2)	7 (1)	14 (3)
You've got ears	7 (1)	7 (1)	-

None of these characteristics makes clear reference to sound travelling; in fact the need for proximity to the sound is the closest that any infants come to considering transmission. In contrast to the clear differences in response pattern between infants and juniors there seems to be very little difference between upper and lower juniors. There is a substantial proportion of children in each of the junior age groups who did not know how sound reached them.

c. *Listening to sounds through an ear trumpet*

Answers to these questions cannot be analysed unambiguously because of the nature of the activity. Children were asked to use a funnel to listen to sounds around them in the classroom and this general instruction made the effects of the funnel less noticeable. Many children reported that the effect of the funnel on their ear was similar to that of a seashell, that they could hear the sea. It was necessary to encourage the child to focus on one particular sound, for example a ticking watch, for the full impact of the funnel to be observed. Some children used the funnel with the large opening to their ear and the small opening facing outwards and this might also have affected responses.

i. What do you think would happen if you had a larger funnel/smaller funnel?

Table 4.5 Children's predictions concerning the effect of different sized funnels upon hearing (Percentages)

	Infants (n = 15)	Lower Juniors (n = 16)	Upper Juniors (n = 21)
Bigger funnel - louder sound Smaller funnel - quieter sound	53 (8)	94 (15)	67 (14)
No effect on sound	13 (2)	6 (1)	24 (5)
Other combinations of funnel size and predicted effect	13 (2)	- -	10 (2)

It is interesting that, having used just one funnel, over half of the children interviewed (and over 90% of lower juniors) felt that a bigger funnel would help them hear better and that a smaller funnel would make the sound quieter ($p < 0.05$). A surprisingly large number of upper junior children (24%) said that the different funnels would make no difference, and this may have been due to the children having difficulty in obtaining the required effect when using the funnel.

This possible explanation is supported by responses to the question about how a funnel affects hearing: 41% of upper juniors said that the funnel made the sound different, and this was a much higher proportion than in the other age groups.

ii. How do you think a funnel affects the way you hear?

This question addressed ideas about sound reception in relation to the ear trumpet.

Table 4.6 Children's ideas about the effect of an ear trumpet on hearing (Percentages)

	Infants (n = 15)	Lower Juniors (n = 16)	Upper Juniors (n = 22)
It makes it sound different	27 (4)	6 (1)	41 (9)
It helps you hear better	7 (1)	25 (4)	23 (5)
It makes the sound louder	20 (3)	6 (1)	9 (2)
It helps sound get into your ears	20 (3)	56 (9)	32 (7)
It lets more air into your ears	-	-	14 (3)
The size and shape of the funnel affect hearing	20 (3)	44 (7)	19 (4)

There is an interesting age difference in responses which children made in terms of the funnel helping them to hear better or making the sound louder. The infant children tended to reply in terms of the sound becoming louder whereas juniors more often thought of the funnel as facilitating the process of hearing. Responses suggesting facilitation also follow this same pattern with children who say the funnel helps sound or air into their ears; fewer infants make this response. This difference between infants and juniors could be indicative of an inability of many younger children to consider sound production-transmission-reception as an integrated process. If younger children consider sounds to be made and to be heard without any need for transmission then there would be no perceived need for the facilitation of picking up sound.

iii. How do you think the sound gets from the classroom to you so you hear it?

Table 4.7 shows that nearly half the infants were unable to explain how sound reaches them and that of the half who did respond a predominant reply is that proximity to the sound is important. Sound appears to be something which happens and can be heard but the process by which this occurs does not seem to be considered by infant children. The ear trumpet does, though, seem to have made children more aware that their ears are the necessary receptor for sound. More mention of the ear was made than with activities such as the rubber band or even the string telephone.

Table 4.7 Children's ideas about how sound reaches them (Percentages)

	Infants (n = 14)	Lower Juniors (n = 16)	Upper Juniors (n = 20)
We're near the sound	14 (2)	19 (3)	-
We're listening	7 (1)	-	-
Sound goes to you	-	13 (2)	15 (3)
Sound goes through holes in the funnel	14 (2)	25 (4)	40 (8)
Sound goes through your ears	14 (2)	31 (5)	35 (7)
Sound goes through the air	-	13 (2)	15 (3)
Other	7 (1)	6 (1)	5 (1)
Don't know	43 (6)	19 (3)	20 (4)

d. Hitting a drum which had rice on the skin

Children were asked to hit a drum upon the head of which was rice, and to watch what happened to the rice. They were asked the following questions:

i. What makes the rice move when you bang the drum?

ii. How do you think the sound gets from the drum to so you hear it?

iii. Do you think you could bang the drum with the rice?

Children's responses to each question will be discussed.

i. What makes the rice move when you bang the drum?

The vast majority of children responded to this question in one of three ways (which were not mutually exclusive). The responses give some indication of children's observation and understanding of vibrations. It cannot be said that their responses relate directly to sound production since the question was worded in terms of movements of the rice. Nearly half of the upper junior sample referred to vibrations or repeated up-and-down movements of the head of the drum. The number of younger children responding in this way was much smaller. Nearly 75% of infant and lower junior children were equally divided between the stick/hit being responsible for the rice moving and the force of the impact of the stick making the head go down then back up once. This latter class of response was clearly identifiable from the children suggesting repeated vibrations; very few ambiguous responses were found.

Table 4.8 Children's ideas about the causes of movement of the rice placed on the drum head (Percentages)

	Infants (n = 14)	Lower Juniors (n = 16)	Upper Juniors (n = 24)
The noise	-	6 (1)	8 (2)
The drum head is tight	-	6 (1)	8 (2)
The bang/the stick	36 (5)	44 (7)	25 (6)
The drum head goes down then up (movement related to hit)	36 (5)	31 (5)	21 (5)
The drum head goes up and down (repeated vibrations)	14 (2)	25 (4)	42 (10)
Air inside the drum pushes the head up	-	6 (1)	8 (2)
Other	7 1	- -	- -

ii. How do you think the sound gets from the drum to you so you can hear it?

Table 4.9 Children's ideas about the transmission of sound from a drum (Percentages)

	Infants (n = 14)	Lower Juniors (n = 16)	Upper Juniors (n = 23)
Characteristic of the drum or the listener	78 (11)	31 (5)	17 (4)
Sound comes out of the drum	7 (1)	25 (4)	52 (12)
Sound goes round inside the drum	-	6 (1)	9 (2)
Sound hits the bottom of the drum	-	6 (1)	13 (3)
Sound goes to you	7 (1)	31 (5)	26 (6)
Sound is carried through the air	-	13 (2)	22 (5)

This question addresses the issues of sound transmission and has some very marked age trends in response. Over three quarters of infant children do not mention transmission but refer to a characteristic of the drum or of the listener, particularly the force used to make the sound, that is, 'you bang it hard' or the existence of ears ($p < 0.001$). The much smaller number of older children who made this type of response tended to mention the force of the hit or the drum being hollow. Table 4.10 shows a more detailed breakdown of this level of response.

The most noticeable trend in Table 4.9 where rate of response increases with age is found with the idea that sound comes out of the drum ($p < 0.02$). Here there is at least a three-fold increase between infants and lower juniors, and lower juniors and upper juniors. Other responses which appear to identify the inside of the drum as the source of the sound refer to sound going round inside the drum or sound hitting the bottom. These three responses show some notion of sound travelling, either within the drum or to leave the drum. A substantial number of juniors (nearly half) refer to sound travelling either through the air or without elaborating further.

Table 4.10 Children's ideas about sound transmission involving characteristics of the drum or listener (Percentages)

	Infants (n = 14)	Lower Juniors (n = 16)	Upper Juniors (n = 23)
You bang it hard	36 (5)	6 (1)	13 (3)
You've got ears	21 (3)	6 (1)	-
I listen carefully	14 (2)	-	-
I'm near it	-	6 (1)	-
It's hollow	-	13 (2)	4 (1)

iii. Can you hit the drum without moving the rice?

This question addresses directly the concept that sound production is associated with movement and some age-related patterns appear in the responses.

Table 4.11 Children's ideas about the connection between sound production and movement (Percentages)

	Infants (n = 14)	Lower Juniors (n = 16)	Upper Juniors (n = 23)
It is not possible to hit the drum without moving the rice	50 (6)	56 (9)	74 (17)
Yes, hit the drum on the side	-	13 (2)	4 (1)
Yes, no explanation	8 (1)	-	4 (1)
Yes, hit it quietly	25 (3)	25 (4)	9 (2)
Yes, other	8 (1)	-	-
Don't know	8 (1)	6 (1)	4 (1)

Half of the infants and lower juniors and three quarters of upper juniors said that it was not possible to hit the drum without moving the rice. If the children who suggested hitting the drum on the sides are added to this, then the number of lower juniors making some association between sound production and movement increases substantially, leaving very little difference between upper and lower juniors. However, the number of upper juniors who think that, were the drum to be hit quietly, the rice would not move is much less than the corresponding number of younger children. This distinction might be a more accurate reflection of the number of children who associate movement of the drum head with vibration and thus with sound production rather than associating movement with the force needed to cause sound production.

e. Listening to everyday sounds

Children were asked about the sounds they could hear around them.

i. How can you hear the sounds around?
ii. What makes a difference to how well you hear?

i. How can you hear the sounds around?

Responses fell into four main categories.

Table 4.12 Children's ideas about how everyday sounds are heard (Percentages)

	Infants (n = 16)	Lower Juniors (n = 16)	Upper Juniors (n = 21)
Characteristic of the listener, or of the force needed for production of the sound	31 (5)	19 (3)	15 (3)
Through your ears	25 (4)	50 (8)	24 (5)
Through holes in windows/ through thin barriers	12 (2)	13 (2)	10 (2)
Through the air	-	13 (2)	14 (3)
Don't know	31 (5)	12 (2)	38 (8)

Approximately two thirds of infants answered this question in a manner which showed no evidence of an idea that sound travelled at all. One third of the infants explained hearing in terms of a characteristic of the listener or of the force of production, and one third answered that they did not know how they heard sound. Substantially fewer juniors responded in this way ($p < 0.05$). Half the lower juniors specified that their ears were the way they heard sounds from around. This category, while not mutually exclusive of any other response does not in itself imply that the child is aware of the transmission of sound. Of the two remaining categories, both imply transmission and there is an interesting difference between them. A small number of children, equally distributed across age bands, suggested that sounds reach them by getting through thin barriers, such as windows, or by coming through holes such as cracks round doors, or keyholes. These children seem to have a notion that sound travels but not that it travels through a medium; in fact, it seems to need the lack of a medium in order to move from sound source to hearer. A similar number of lower and upper juniors (but no infants) mention sound travelling through air.

ii. What makes a difference to how well you hear?

Children were asked to any conditions they thought would affect their hearing. This question was found by interviewers to be difficult to phrase successfully and this problem might explain the high number of upper juniors who could not answer.

Table 4.13 Children's ideas about conditions which would affect their hearing (Percentages)

		Infants (n = 15)	Lower Juniors (n = 16)	Upper Juniors (n = 22)
Nothing		20 (3)	19 (3)	-
Volume		47 (7)	50 (8)	23 (5)
Pitch		-	-	5 (1)
Distance		47 (7)	38 (6)	41 (9)
Obstacles		7 (1)	-	9 (2)
Background noise		40 (6)	13 (2)	32 (7)
Concentration on sound		-	-	18 (4)
Looking		-	-	5 (1)
Number of conditions mentioned	0	3	5	-
	1	5	5	10
	2	2	4	6
	3	4	1	1
	4	-	-	1

Interestingly, an equal percentage of infant and lower junior children thought that nothing could affect their hearing. All upper juniors mentioned at least one condition. The majority of infant responses were concerned with volume and distance, as well as background noise. The relatively high percentage of infants compared with juniors suggesting this latter condition might be due to the open plan design of one of the schools. Every child interviewed from that infant class mentioned background noise, and no other infants gave that response. This experience could certainly have coloured their observations of sound.

A comparison between the pre-Intervention sound activities

Each of these activities made a particular aspect of sound more explicit than others, and because of this context-related issue the questions which were asked about each activity were specific to that one. However, it is possible to compare the different activities by using a number of super-ordinate categories which are common to all the activities. These categories are concerned with:

a. sound production

b. sound transmission

c. sound reception

a. Sound production

Children's ideas about how sounds were made seem to fall into three main groups:

i. Some children gave explanations which related to physical properties of the sound-making object, or made tautological statements about sound production. There was no mention of any action being associated with sound-making.

e.g. Q *Where was that noise coming from?*

R *Off the rubber band.*

Q *How was it doing that?*

R *It stretched.*

Q *What do you think made the band make a noise?*

R *Because it's an elastic band and it's bouncy.*

ii. Other children's explanations did made an association between an action/movement and sound production. Mention of talking was included in this category.

e.g. Q *What do you think makes the drum make a noise?*

R *It makes a noise when you bang it*

Q *What makes the rubber band make a sound?*

R *You pull it and it hits the other side and makes a sound*

iii. Some children answered in a way that suggested that sound production was linked to the vibrations of an object.

e.g. Q *What do you think makes the drum make a sound?*

R *You bang it - the skin vibrates and the sound goes round inside and then comes out.*

Q *Do you think you could tell if the rubber band was making a sound if you couldn't hear it?*

R *You'd see it shaking. When it shakes it always makes a sound.*

The production of sound was the most prominent aspect of two of the activities, the rubber band and the drum, and children were questioned directly about sound production in those contexts: with regard to the string telephone, the ear trumpet and the everyday sounds children's mention of sound production as part of their answers to other questions was scored.

Table 4.14 A comparison of the number of children holding particular ideas about sound production across activities (Percentages)

(infant, n = 13; lower junior, n = 14; upper junior, n = 17)

	String telephone			Rubber band			Ear trumpet			Drum			Everyday sounds		
	Inf	LJ	UJ	Inf	LJ	UJ	Inf	LJ	UJ	Inf	LJ	UJ	Inf	LJ	UJ
No mention of action associated with sound production	38 (6)	29 (6)	12 (6)	31 (4)	-	6 (1)	85 (11)	50 (7)	82 (14)	15 (2)	-	-	77 (10)	-	18 (3)
Some mention of action associated with sound production	62 (8)	57 (8)	65 (11)	69 (9)	50 (7)	41 (7)	15 (2)	50 (7)	18 (3)	85 (11)	79 (11)	53 (9)	23 (3)	100 (14)	82 (14)
Sound production associated with vibration	-	14 (2)	24 (4)	50 (7)	53 (9)	-	-	-	-	21 (3)	40 (8)	-	-	-	-

Across all activities, with the exception of the ear trumpet, there is a trend towards more scientific ideas related to increasing age, and the main differences seem to be between infants and juniors. No infants mentioned an association between sound production and vibrations and a substantial number did not mention a link between any action and sound production.

The vast majority of responses to each of the four activities (excluding the ear trumpet) fell into the category of some action being associated with sound production. The ear trumpet and everyday sounds both gave children less opportunity to mention sound production, particularly the ear trumpet. These two activities both focused upon sound reception and it was not necessary for children to identify and mention a particular localised sound source in order to answer the questions. The data should not be interpreted in terms of context specificity but in terms of differing opportunities to respond during the interview.

As might have been expected, the two activities in which movement was observable as the result of an action (the drum and the rubber band) led to the most mention of vibrations. Interestingly, the string telephone also elicited similar responses. It is possible that, since a medium for transmission ws explicit, children were more likely to speculate about how that pathway could be used. Children might also have felt the yogurt cups vibrating as they spoke into them.

The notion of vibration

It was felt that any mechanism which was associated with sound production could be made apparent by using a means of categorising the information which would draw out the perceived relationships between sound production and vibrations. Children's responses which either mentioned the word 'vibration' (whether it was in connection with sound production, transmission or reception) or used a synonym, for example 'wobble', were all included in this analysis.

Table 4.15 Children's ideas about the perceived link between vibration and sound production (Percentages)

(lower junior, n = 14; upper junior, n = 17)

	String telephone		**Rubber band**		**Ear trumpet**		**Drum**		**Everyday sounds**	
	LJ	UJ	LJ	UJ	LJ	UJ	LJ	UJ	LJ	UJ
Sound cause vibrations	21 (3)	41 (7)	-	-	-	-	7 (1)	35 (6)	7 (1)	-
Vibrations cause sound	-	-	43 (6)	29 (5)			21 (3)	24 (4)	7 (1)	-
Sound is vibration	-	6 (1)	-	12 (2)	-	-	-	-	-	-
Unclear/ inconsistent use of word	-	12 (2)	7 (1)	6 (1)	-	-	-	6 (1)	-	-
Vibrations not mentioned	79 (11)	41 (7)	50 (7)	53 (9)	100 (14)	41 (7)	71 (10)	35 (6)	86 (12)	100 (17)

There are interesting differences in responses here which might suggest that children are considering mechanisms of sound production which are specific to particular activities. Children who mentioned vibrations in connection with the string telephone suggested that the sound caused the vibrations, whereas the opposite pattern was found with the rubber band where the vibrations were thought to cause the sound. The drum, on the other hand, elicited both of these mechanisms. It is possible that the string telephone might be a special case since the sound being produced is the child's own voice which children might not consider to be an example of their definition of sound. The vibrations in the pharynx, which are easily felt, might thus be associated with voice production, as distinct from sound production. The vibrations could be an attempt to explain voice transmission rather than sound production. The rubber band might lead children to suggest that the vibrations cause the sound because of the nature of their observations while plucking the band. The children stretch the band and let it go before any sound is heard, so the larger movements, causing band to be stretched and let go, might be observed rather than the smaller vibrations. The drum head, on the other hand, is stationary until the stick hits it and the moment that the stick hits the drum appears to be exactly the same as the moment at which the sound is produced. These simultaneous events might make the choice of mechanism less evident from observation. These could have led to both being suggested with approximately equal frequency.

b. Sound transmission

Children had a variety of ideas about how sound travelled and they can be put into three main groups:

i. Some children made no mentioned at all of sound travelling. They explained being able to hear a distant sound in one of several ways; either in terms of the force used to produce the sound, the proximity of the sound, its volume, or as a characteristic of the listener.

e.g. Q *How do you think the sound gets to you so you can hear it?*

R *Because I've got ears and I can listen.*

R *Because you make it bang*

R *Because it's loud*

ii. Another group of children had some idea that sound travelled, but seemed to have no notion about the medium through which sound passed. Some of these responses suggest that sound can only travel without a medium, i.e. that it will only go through holes.

e.g. R *Tunes are very small and they can get through the gaps in the doors*

R *The sound comes to your ear; I don't know how*

iii. Some children seemed to know that sound travelled through a medium.

e.g. R *The message goes through the stretched wire to your ear*

R *The air brings the sound up to your ear.*

Table 4.16 A comparison of the number of children holding particular ideas about sound transmission across activities (Percentages)

(infant, n = 13; lower junior, n = 14; upper junior, n = 17)

	String telephone			**Rubber band**			**Ear trumpet**			**Drum**			**Everyday sounds**		
	Inf	LJ	UJ	Inf	LJ	UJ	Inf	LJ	UJ	Inf	LJ	UJ	Inf	LJ	UJ
No mention of sound travelling	46 (6)	21 (3)	-	100 (13)	43 (6)	53 (9)	77 (10)	29 (4)	35 (6)	77 (10)	36 (5)	47 (8)	100 (13)	57 (8)	76 (13)
Some mention of travel, no mention of medium	7 (1)	7 (1)	6 (1)	-	43 (6)	29 (5)	23 (3)	57 (8)	47 (8)	23 (3)	50 (7)	41 (7)	-	29 (4)	6 (1)
Mention of sound travelling through a medium	46 (6)	71 (10)	94 (16)	-	14 (2)	18 (3)	-	14 (2)	18 (3)	-	14 (2)	12 (2)	-	14 (2)	18 (3)

Sound transmission through a medium was made explicit in only one activity, the string telephone. Children spoke into one yogurt pot and the message was received through the other, the observable connection between the two pots being a piece of string. It can be seen from Table 4.16 that the number of children mentioning that sound travelled through a medium was substantially greater for each age group when talking about the string telephone than about any of the other activities. Between activities, the largest differences were found with the upper junior sample where they were significant with respect to each activity: rubber band ($p < 0.01$), ear trumpet ($p < 0.05$), drum ($p < 0.001$) and everyday sounds ($p < 0.001$). There were also differences in infant responses between the string telephone and the rubber band ($p < 0.01$) and everyday sound ($p < 0.01$), though the lower junior responses were not significantly different than would have been expected.

A comparison, between activities, of numbers of children who mentioned that sound travelled shows a consistent pattern of responding, with the exception of the string telephone. This pattern shows fewer infants than juniors mentioning travel, and also very little difference between the lower and upper junior samples. The majority of children at each age also make no reference to a medium through which sound travels (with a non-significant exception for upper juniors concerning everyday sounds). This pattern suggests that the majority of children have little or no notion of sound travelling through a medium. The example of the string telephone, where the medium is visible, encourages far more children to mention a medium, but this response seems to be specific to that particular context.

Another angle which should be considered is whether the children who mentioned sound travelling through air regarded air as a medium, as something with a volume, mass and density, or whether they were giving the accepted name to the empty space around. Children made very few attempts to explain sound transmission.

Three quarters of the infant children who mentioned that sound travelled could give no further explanation. This number fell to around half in the upper and lower junior samples. The idea that sound travelled as words was more prevalent in infant children. This representation could be due to the inability of these infants to abstract the notion of 'sound' from that of the message which they had spoken or received. As might be expected, the use of words which tend to be part of a scientific vocabulary was restricted to junior children, with nearly half the upper junior sample referring to vibrations. The use of the word, 'vibration', does not necessarily imply that the term is being used in a scientifically acceptable way. When asked for clarification, some children talked about 'wobbling' or 'moving up and down' while others could not define the term at all.

Table 4.17 A comparison of ideas about how sound travels across activities (Percentages)

(infant, n = 13; lower junior, n = 14; upper junior, n = 17)

	String telephone			**Rubber band**			**Ear trumpet**			**Drum**			**Everyday sounds**		
	Inf	LJ	UJ	Inf	LJ	UJ	Inf	LJ	UJ	Inf	LJ	UJ	Inf	LJ	UJ
Sound travels, no further explanation	38 (5)	43 (6)	65 (11)	-	57 (8)	29 (5)	23 (3)	64 (9)	65 (11)	23 (3)	50 (7)	35 (6)	-	36 (5)	24 (4)
Sound travels as words	15 (2)	14 (2)	6 (1)	-	-	-	-	7 (1)	-	-	-	-	-	7 (1)	-
Sound travels as sound waves*	-	7 (1)	-	-	-	6 (1)	-	-	-	-	-	6 (1)	-	-	-
Sound travels as vibrations*	-	14 (2)	47 (8)	-	-	12 (2)	-	-	-	-	-	12 (2)	-	-	-
Sound travels as echoes*	-	-	6 (1)	-	-	-	-	-	-	-	14 (2)	-	-	-	-
No mention of sound travelling	46 (6)	21 (3)	-	100 (13)	43 (6)	53 (9)	77 (10)	29 (4)	35 (6)	77 (10)	36 (5)	35 (6)	100 (13)	57 (8)	76 (13)

*child has actually used the word

The use of 'sound waves', though restricted to a very small number of children, seemed to be associated with an interesting model of sound transmission. The sound waves appeared to be present in the air, possibly like waves on the sea, and to pick up the sound when it was produced. An analogy could be that sound waves were like buses and they picked sound up, carried it and put the sound down again. Whether the sound waves were thought to float in the air, or to be air in a certain form, was not clear from the children's comments.

c. *Sound reception*

There seemed to be a smaller range of ideas concerning reception though again they can be divided into three groups.

i. Some children made no mention of a receptor, refering simply to 'hearing'.

e.g. Q *How can you hear the sound from the drum?*

R *You can hear it*

ii. Some children specified that the ear is necessary for sounds to be heard.

e.g. R *The sound goes to your ears and you hear.*

iii. A few children thought that the ear drum had to vibrate for sound to be heard.

e.g. *There's an ear drum in your ear and every time something makes a noise it's like it's banging on your ear drum.*

Table 4.18 A comparison of the number of children holding particular ideas about sound reception across activities (Percentages)

(infant, n = 13; lower junior, n = 14; upper junior, n = 17)

	String telephone			Rubber band			Ear trumpet			Drum			Everyday sounds		
	Inf	LJ	UJ	Inf	LJ	UJ	Inf	LJ	UJ	Inf	LJ	UJ	Inf	LJ	UJ
No mention of a receptor	92 (12)	64 (9)	76 (13)	100 (13)	36 (5)	71 (12)	62 (8)	7 (1)	6 (1)	62 (8)	36 (5)	65 (11)	77 (10)	50 (7)	76 (13)
Mention of the ear as the receptor	7 (1)	36 (5)	24 (4)	-	57 (8)	29 (5)	38 (5)	93 (13)	35 (6)	38 (5)	57 (8)	35 (6)	23 (3)	43 (6)	24 (4)
Mention of vibrations being set up in the receptor	-	-	-	-	7 (1)	-	-	-	-	-	7 (1)	-	-	7 (1)	-

The main activity to emphasise sound reception was the ear trumpet, and an inspection of Table 4.18 shows that the main significant differences in response are concerned with junior children and the ear trumpet. It is perhaps surprising that the string telephone did not elicit more reference to ears. This lack of mention might be due to the sounds involved in the string telephone activity being speech, and speech being considered to be distinct from sound. This explanation would be consistent with the interpretation of vibrations as a mechanism of voice transmission rather than when mentioned in connection with the string telephone.

The three mentions of vibrations being set up in the receptor were all from the same eight-year-old boy. His responses did not in fact refer to vibrations but to 'sound banging on the eardrum', a notion which could be a direct analogy with producing sound on a drum by hitting it with a stick. (Mention of the ear drum was not uncommon but this case is the only one where an attempt was made to explain its function.)

A small number of children in each age band mentioned listening either as a part of all of or their explanation of how they could hear sounds. These children considered their attending to the sound to be a necessary condition for hearing sounds.

Table 4.19 Prevalence of the notion of Active Listener (Percentages)

	(infant, n = 13; lower junior, n = 14; upper junior, n = 17)				
	String telephone	**Rubber band**	**Ear trumpet**	**Drum**	**Everyday sounds**
	Inf LJ UJ	Inf LJ UJ	Inf LJ UJ	Inf LJ UJ	Inf LJ UJ
Active listening	23 14 18 (3) (2)(3)	7 - - (1)	7 14 - (1) (2)	31 7 - (4) (1)	15 7 12 (2) (1)(2)

These 'active listeners' each tended to give this explanation for only one activity; one infant girl, however, used this psychological model for every activity, and one lower junior girl applied it to the string telephone, everyday sounds and ear trumpet which are the three activities making most reference to sound reception.

4. INTERVENTION

The Exploration phase encouraged children to express a wide range of ideas about sound both to the Project team, during individual interviews, and to their teachers, through less intensive elicitation work in the classroom. These ideas were the foundation upon which the classroom Intervention was built.

The Project team had developed a range of Intervention strategies as a result of ideas expressed by children during the first round of topics. The teachers, and children, were thus familiar with using these strategies in the classroom and it was suggested that they applied them to their work on 'Sound'. These strategies were designed to help teachers structure activities in four areas, each of which appeared to be capable of exerting important influences on the manner in which children form their ideas.

The teachers met for one day prior to the Intervention to allow them to discuss with each other and the Project team the manner in which they would implement the Intervention. During the Intervention, there were two half-day meetings; one towards the middle and one towards the end of the five-week Intervention period. These meetings enabled teachers to share experiences and pool ideas and were very valuable for the support they provided. The classroom Intervention took place between April 25th, 1988 and May 27th, 1988, the majority of the first half of the summer term.

Teachers were asked to enable their classes to engage with at least one activity relating to each strategy during the Intervention phase. The information which teachers were given concerning the Intervention phase may be found in Appendix V.

4.1 Intervention Strategies

The following paragraphs describe the four strategies and the manner in which they could be appropriate for the sound work.

i. Helping children to test their own ideas

Children's ideas about sound appeared to be based upon observations they had made, or upon information which they might have acquired from secondary sources, for example teachers, books or television. In either case, the ideas seemed to be very context-specific, relating just to one type of experience or observation. These observations were often incomplete or misinterpreted, leading children to reach conclusions which would not be substantiated by reference to wider contexts or by more systematic observation. It was envisaged that, by encouraging children to formulate questions which they could test in a systematic way, children might be enabled to develop their thinking along more productive lines. This strategy was the main focus of each class's Intervention experiences.

Fig. 4.1

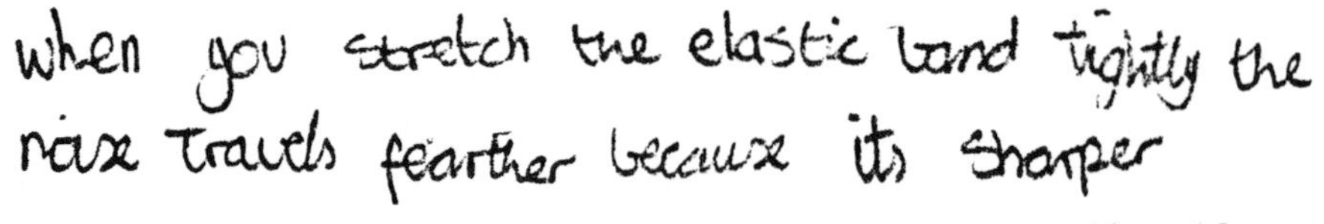

when you stretch the elastic band tightly the naise travels fearther because its sharper

(Age 10 years)

"When you stretch the elastic band tightly the noise travels farther because it's sharper"

Fig. 4.1 gives an example of an observation which was made, and from which an inaccurate conclusion had been drawn. While it might not be very easy to test whether sound travels further when it is higher in pitch, a carefully constructed test might encourage the child to draw an alternative conclusion from the evidence which could be collected.

The role of the teacher was very important during such investigations and the teachers were required to develop a degree of competence in questioning children and in process science. It was through encouraging children to pose a question which was investigable, and by helping children to increase their awareness of the need to conduct rigorous scientific tests that teachers could encourage children to challenge their original notion. The pupils were being asked to put their ideas to the test and regard them as statements which could be discarded if they were found to be inadequate. These skills and attitudes, for both the pupils and the teachers, needed to be developed within the framework of a secure, open learning environment, over a period of time. The success of the investigations in helping the children to challenge their ideas might have been influenced by the extent to which the above-mentioned factors had been developed prior to and during the Project work.

ii. Encouraging children to generalise from one specific context to others through discussion

A recurrent theme throughout Section 3 was the high degree of context specificity apparent in children's responses. The same child would often give an answer that related to observations made for one activity and, because different observations had been made for another activity, a different explanation would be given.

Fig. 4.2

(Age 7 years)

"I think the drum makes a sound because it is hollow and it echoes"

It was felt that, if the teacher provided opportunities for children to discuss their observations and encouraged them to see areas of commonality between classroom activities and their everyday experiences then children might develop concepts which were broader and less context-specific. By participating in class discussions children would be able to interact with the views and ideas which were held by their peers and this could enable children to broaden their experiences. This approach to class discussion required some teachers to reappraise their definitionof the activity. In many instances discussions had been considered to be question and answer sessions led by the teacher's questions. Over the course of the Intervention some teachers reported that their role in class discussions was changing: they found themselves to be chairing discussions between pupils. The pupils themselves were generating examples and instances from which to draw their own generalizations, within the limits of the discussion as defined by the teacher. These limits specified the general area within which the discussion should range and were defined in order to ensure that the exchange of ideas was focused enough to be profitable, and for the children to be able to see the common ground between each other's experiences.

iii Encouraging children to develop more specific definitions for particular key words

There were several instances of children using words for which there was an accepted definition in an idiosyncratic manner. 'Vibrate' and 'echo', for example, were each used on some occasions to mean 'repeat'.

Fig. 4.3

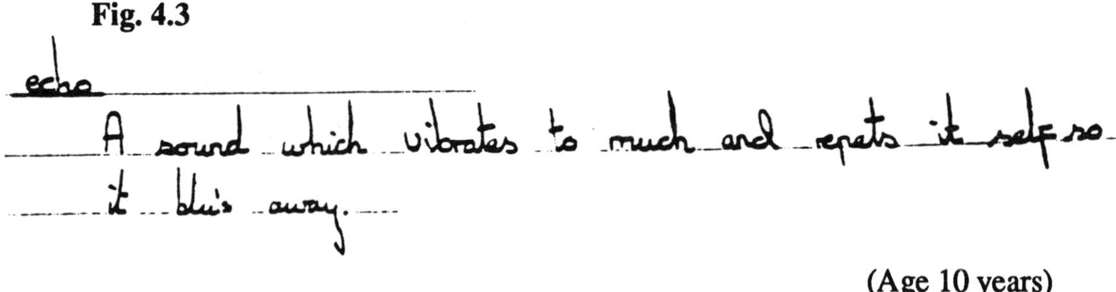

(Age 10 years)

"*echo A sound which vibrates too much and repeats itself so it blows away*"

Fig. 4.4

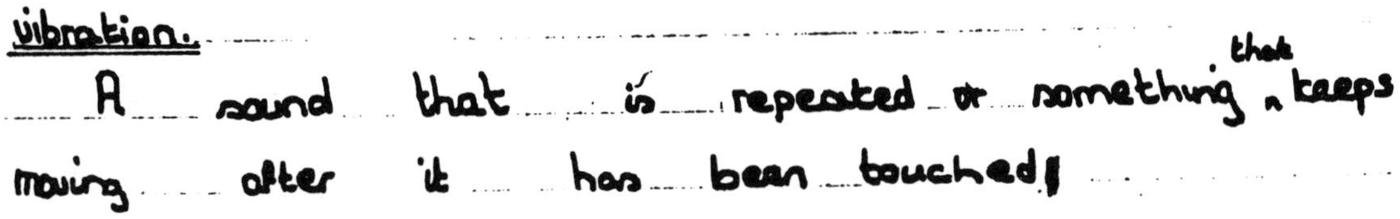

vibration.
A sound that is repeated or something that keeps
moving after it has been touched.

(Age 10 years)

"*vibration A sound that is repeated or something that keeps moving after it has been touched*"

This usage suggested that children had incorporated one aspect of the scientific definition into their vocabulary, possibly the one which had been observed in the context in which they first heard the word. This aspect of the word might then have been embellished by the child so that a definition related to, though not synonymous with, the dictionary would be developed and used. It was considered important that teachers clarified the meanings attached to these important words by children, and encouraged the children to reach consensus definitions by using the words in a relevant context. The emphasis of this vocabulary work was intended to be on using the words as part of an activity, so that the context was explicit and common to everyone.

In practice, the generalization and vocabulary activities were often very closely related. Where children were using words in an idiosyncratic way it was often helpful to encourage children to generalize as a means of clarifying their definition of the word. It was envisaged that such activities would clarify vocabulary which the children were already using rather than introduce unfamiliar terms. However, it was acknowledged that this latter approach could be appropriate if children obviously lacked a label for a specific process which they could describe.

iv. Finding ways to make imperceptible change perceptible

Sound is perceptible through the sense of hearing but it is not visible to the eye, except as vibrations in some sound producers. The process of sound transmission, in particular, is difficult to perceive. There were several activites which made sound visible (or tangible), for example placing rice on the drum or using a string telephone, and it was felt that the children could profitably explore these avenues.

The potential value of encouraging children to develop a form of representation for sound on paper was also considered, and was attempted by some teachers.

4.2 Preparation for Classroom Intervention

At the teacher meeting prior to the Intervention the teachers were grouped so that teachers who taught the same age children were together: thus infant teachers formed one group, lower junior teachers another and upper junior teachers a third. In these groups the teachers decided upon the open question they would present to children to begin the Intervention. This question was specific to each age group and ensured that the children's investigations would have the same starting point. The question was not intended to restrict children to one particular investigation but to provide a framework which would assist children in posing an investigable question. For example, "What do you think affects the way you hear?" could be a starting question which would enable children to suggest a particular variable to investigate.

The question decided upon by each group of teachers was very similar to the one exemplified above, though worded differently for children of different ages.

Infants - How can we make out ears better to hear with?

Lower juniors - What sort of things can alter the way you hear the sounds?

Upper juniors - What sort of things affect how well you can hear?

Teachers were asked to collate their class Intervention work on a simple record sheet, noting the activity with which children were engaged, e.g. the drum; the question which they posed to begin discussion about the investigation; the investigation which ensued; related vocabulary activities, and generalizations which were encouraged through discussion.

4.3 *Classroom Implementation of the Intervention*

Infants

i. Helping children to test their own ideas

The investigations which were carried out by infant children were based upon the reception of sound. Children in one class made large paper ears and devised a way of fixing them round their own ears. Another class used different shapes and sizes of funnels and cylinders to see how these affected hearing. These hearing aids were all tried out using a standard sound. The distance of the child from the sound was also controlled: it was either kept constant, so that a subjective judgement of perceived loudness with and without the 'big ear' was obtained, or it was gradually increased until the noise was too quiet to hear.

One classroom happened to have a squared floor and the teacher encouraged the children to use the squares as the measure of distance. Some children were able to transfer their results onto squared paper. The diagram in Fig. 4.5 shows how well the child could hear a ticking clock with a funnel to his ear. The number of ticks indicated the perceived loudness of the clock.

Fig. 4.5

1	2	3	4	5	6	7	+	Funnel
1	2	3	4	5	6	✓		
1	2	3	4	5	x			
1	2	3	4	✓				
1	2	3	✓✓					
1	2	✓✓✓						
1	✓✓✓✓							

(Age 6 years)

The same class also compared different shapes and sizes of funnels and tubes. This investigation gave inconclusive results and the children were unable to suggest any reasons why the different shapes of the funnels had different effects on the volume of the sound which was heard.

ii. Generalizations and vocabulary

During the course of the investigations it became apparent to some of the teachers that many children were associating the size of an object with the volume of sound which it produced (Fig. 4.6).

Fig. 4.6

(Age 6 years)

Following discussions about loud and soft during which teachers asked children to give examples of loud sounds and soft sounds from objects of varying sizes, children again drew loud and soft sounds. There was still a high degree of correlation between size and perceived volume of noise, but some objects were depicted which were either large and made a quiet noise (Fig. 4.7) or were smaller and made a loud noise (Fig. 4.8).

Fig. 4.7

soft loud

(Age 6 years)

Fig. 4.8

loud loud

quiet sounds
watch

(Age 6 years)

Lower Juniors

i. Helping children to test their ideas

The lower junior class teachers introduced the Intervention period to their classes by using the stimulus of a clock ticking in a box and asking the children how they thought they could alter the way they heard the clock. The children then generated lists of variables which they thought would affect their hearing of the clock (Fig. 4.9).

Fig. 4.9

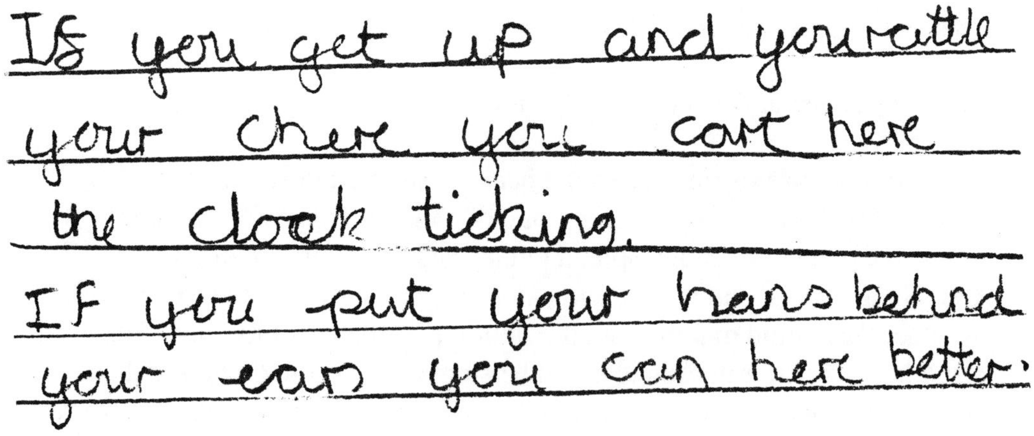

If you get up and you rattle
your chare you cant here
the clock ticking.
IF you put your hans behnd
your ears you can here better:

(Age 8 years)

> *"If you get up and you rattle your chair you can't hear the clock ticking. If you put your hands behind your ears you can hear better"*

The children selected a variable to investigate from the list. The investigations concerned the effects on hearing of:

. proximity to the sound source

. blocking sound transmission at the site of production or the site of reception

. improving hearing using a funnel or other aid

Emphasis was placed upon the children conducting well-controlled tests. The children devised various methods of making a noise which could be exactly replicated in terms of volume, pitch and duration. Some methods used were an electronic keyboard, and dropping an object from a pre-determined height. Children also either kept a consistent distance from the sound source or used distance as the variable which was measured. Some children required a lot of help in planning and constructing a fair test but it tended to be matters of practicalityrather than of principle which were problematic. The recognition of the necessity to control volume and proximity suggested that children of this age understood the significance of these two variables for hearing. Children who were investigating the effects of proximity on hearing might therefore have been confirming a notion which they assumed to be true rather than testing an idea of which they were unsure.

Proximity

Children decided that the further away they were from the constant sound source the harder it was for them to hear. Some children could hear the sound from much further away than others and it was suggested that the levels of background noise had been very variable and that this factor was likely to have affected the results.

Blocking sound at the site of production

In attempting to explain why sound became quieter the further the listener was from the source one group of children suggested that the sound went into the air and spread out around the room. In order to test whether the sound was moving and escaping they decided to find out whether they could trap the sound by putting obstacles in its way. They hypothesised that if the sound could be stopped by a sound blocker then what normally happened to the sound would be that it did spread out from the sound source. The children predicted that the sound would spread out and that they would be able to block the sound from a radio with one of a range of materials placed over the speaker (Fig. 4.10).

Fig. 4.10

(Age 7 years)

The children found that the sound could still be heard through a range of materials.

Fig. 4.11 first we used thin meterial then we put more layers of meterial to make it thicker and see if we can hear it we have used a carrier bag we put the carrier bag were the sound was coming out and pressed a button.

Material	what happend
thin carrier	We could hear through the thin carrier
thick carrier	✓
silk carrier	✓
and hand.	✓
Class door	✓
class	✓

(Age 7 years)

The teacher encouraged them to try an alternative avenue to challenge their conclusion that the sound could be heard because it came through thin layers.

Transcript 4.1

"Child 1 *We folded it until you couldn't fold it any more. We would still hear it only slightly.*

T. *Why do you think you could still hear it?*

Child 2 *Because it's thin.*
Child 1 *It was coming through. I thought it wouldn't have done.*
Child 3 *It came through the silk even when hand was over it.*
Child 1 *Even a quiet noise.*

T. *Can you try anything else?*

Child 2 *Try all. Hand + material + carrier.*
Child 3 *Sheep's wool + silk + hand + carrier. you can hear it.*

T. *Will it come through anything?*

Child 2 *It won't come through anything.*

Child 2	*Try to find something it won't come through.*
T.	*Why can we hear the recorders now?*
Child 3	*The door's open.*
Child 1	*No, it's closed - it's louder. We'll try the door.*
Child 2	*It's got a key hole.*
Child 3	*Cover the key hole. We can hear it.*
Child 2	*There's a crack at the top and the bottom - somebody lie on the floor. There's a crack at the top - it won't go through there - it would have to go up and then down.*
Child 3	*I can hear it. It could come through the door and my ears get cold.''*

Having blocked every hole around the classroom door the sound was still audible.

Fig. 4.12

(Age 7 years)

The children asserted that sound could travel through anything, thick or thin, though particularly around the edges. This investigation had challenged their notion about the media through which sound could travel, though the children appeared to be reluctant to incorporate their findings into their ideas. Instead they suggested that the sound must have found holes or edges through which to travel. This notion that sound must travel unimpeded through a space has been mentioned in Section 3 and was most commonly found among children of lower junior age.

Blocking sound at the site of reception

Children from another class decided to investigate the efficacy of different materials in ear muffs. The materials were placed inside yogurt cartons of the same size and held in place over the ears. A ticking clock was used as the sound source.

Fig. 4.13

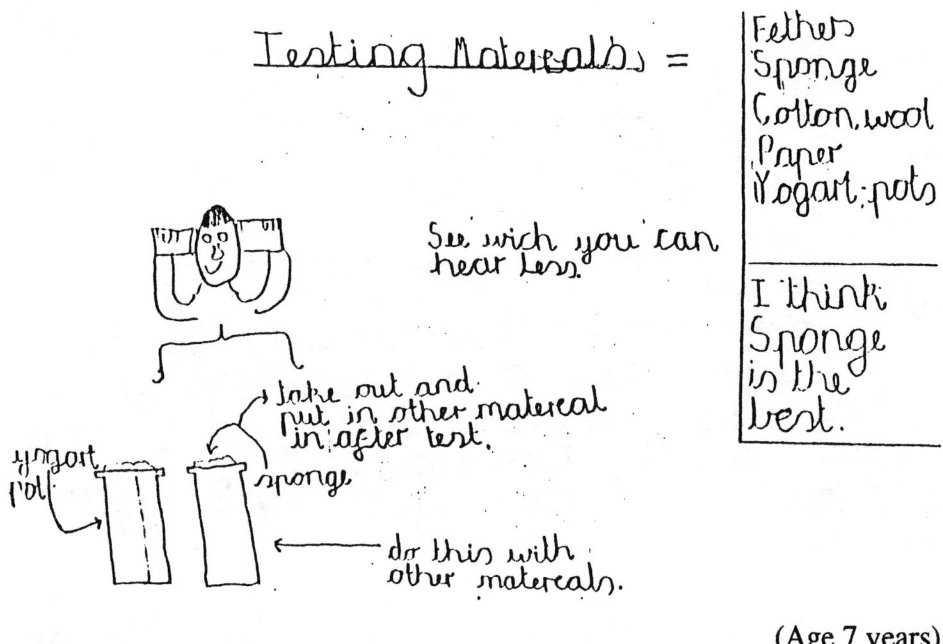

(Age 7 years)

Most children found that the cotton wool ear muffs were the best, and one boy's conclusions were that the cotton wool 'absorbed and trapped the sound waves'.

Improving hearing

Children used funnels of different sizes to listen to sounds. They predicted that any funnel would result in improved hearing, and that a bigger funnel would be better than a smaller one. They found that the bigger funnel was more effective as a hearing aid and decided that this was because more sound could get into the larger funnel.

ii. Encouraging children to develop more specific definitions for particular key words

Two words which were used by a large number of lower juniors, and which teachers felt needed clarification, were 'echo' and 'vibrate'.

Children from one class were asked to 'Draw a picture to show how you think you get an echo. Show where the sound comes from and what happens to it'. Two main ideas emerged about the meaning of the word: some children drew themselves in an enclosed space and explained that the sound hit the boundaries of the space and bounced back so that the sound was heard again (Fig. 4.14). Other children showed themselves in an unenclosed area and mentioned the sound repeating itself as it moved further away from them (Fig. 4.15). This second explanation did not involve any notion of the sound being deflected and returning to the speaker.

Fig. 4.14

(Age 7 years)

''When you are standing in a tunnel and you shout something it will bounce onto the wall and bounce back to you. And that is how you get your echo.''

Fig. 4.15

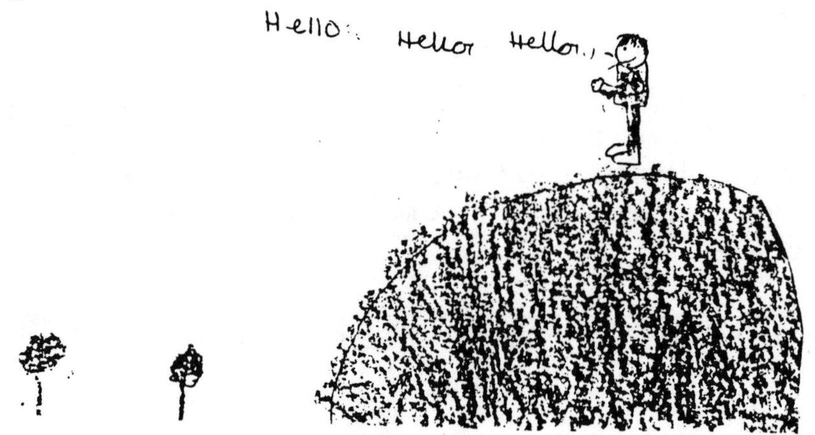

(Age 7 years)

Fig. 4.16

(Age 8 years)

A small number of children appeared to combine these models to present a repeated message which was deflected back to the sender (Fig. 4.16).

This vocabulary activity about echoes was also useful as a starting point for a discussion in which children could share their ideas about the locations in which they had experienced echoes. This discussion revealed some very specific instances of echoes (Fig. 4.17) and enabled some children to broaden their concept of an echo so that it could encompass more than one location.

Fig. 4.17

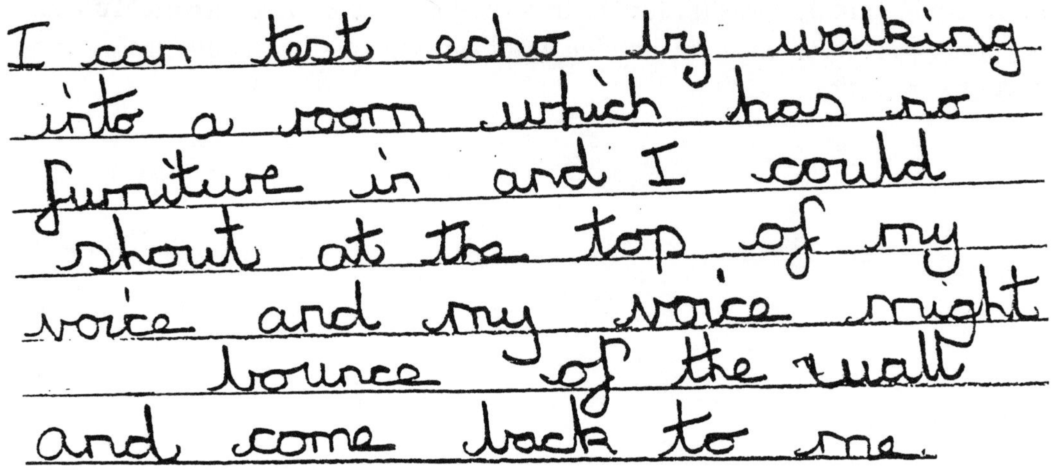

I can test echo by walking
into a room which has no
furniture in and I could
shout at the top of my
voice and my voice might
bounce of the wall
and come back to me.

Upper Juniors

i. Helping children to test their ideas

The children in the upper junior age group found the manipulation and control of variables far easier than did younger children. This aspect of the classroom work needed less teacher intervention than with younger children, and the children tended to control their investigations without being prompted.

The range of investigations which children of this age carried out was wider than in the younger two age groups, and it encompassed most of the investigations which the younger children had done. There was a noticeable difference in the understanding which children brought to their investigations and the idea of sound transmission was evident in the design and the recording of the investigations. The majority of the investigations tested ideas about the effects on hearing of one of the following variables:

. distance between hearer and sound source

. blocking sound reception

. background noise

. the positioning of the sound source relative to the hearer

. the existence of different media between hearer and sound source

Investigations involving distance and sound reception were similar to those carried out by younger children. However, they provided children with a need to develop a pictorial representation of sound which would enable them to explain the results they had found (Fig. 4.18). The example shown in Fig. 4.18 shows that this child has depicted sound transmission being affected by the blocking of sound reception. This diagram could form the basis for further Intervention work in which the child could make predictions based upon the drawing, test them and possibly modify the notation and ideas as a result.

Fig. 4.18

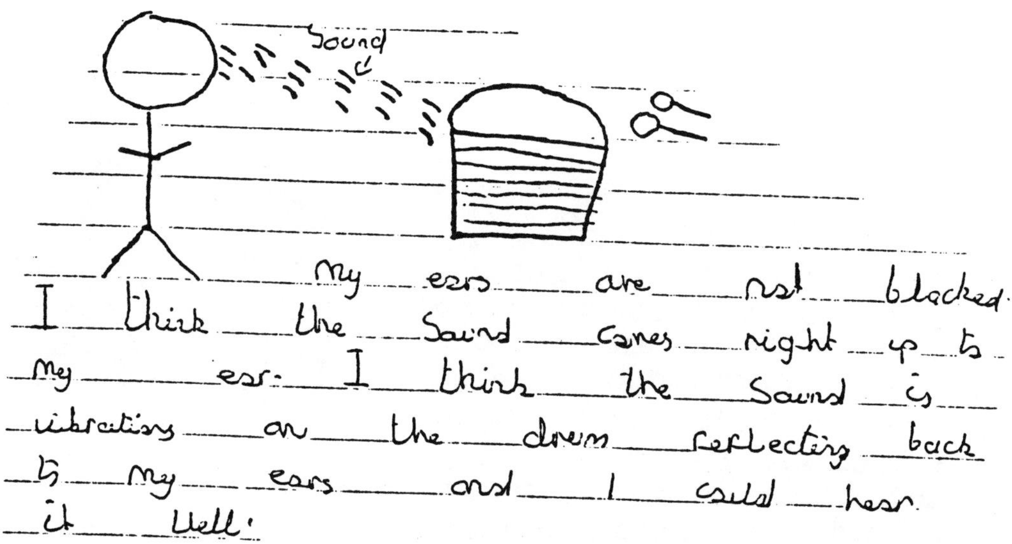

"My ears are not blocked. I think the sound comes right up to my ear. I think the sound is vibrating on the drum reflecting back to my ears and I could hear it well."

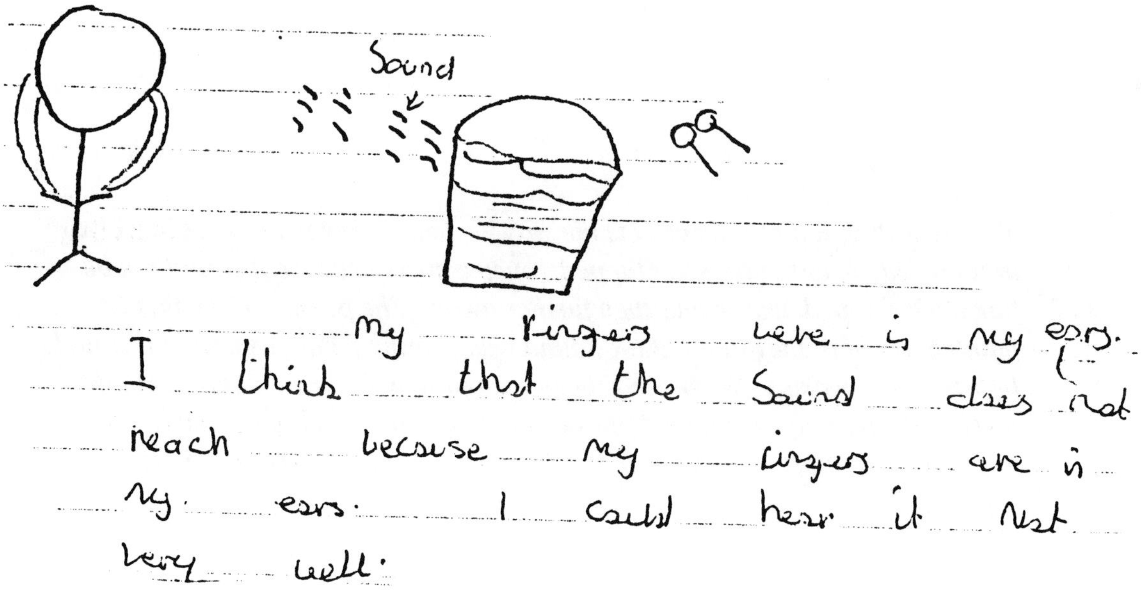

(Age 9 years)

"My fingers were in my ears. I think that the sound does not reach because my fingers are in my ears. I could hear it not very well."

Children who investigated the effect of the direction of the sound source relative to the hearer appeared to be surprised that their test was challenging to carry out (Fig. 4.19) and interpret (Fig. 4.20).

Fig. 4.19

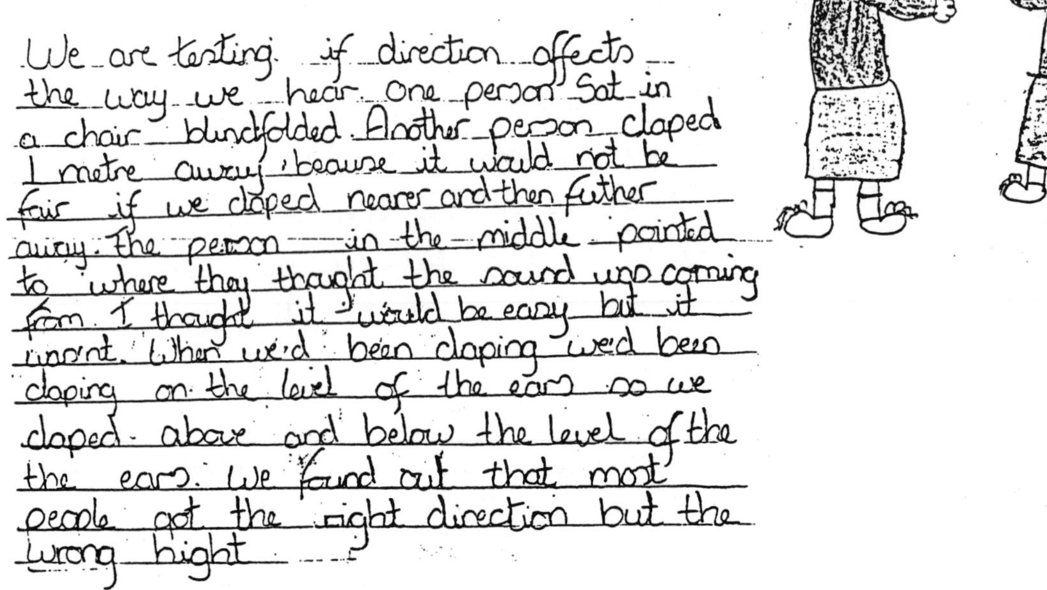

(Age 9 years)

"We are testing if direction affects the way we hear. One person sat in a chair blindfolded. Another person clapped 1 metre away because it would not be fair if we clapped nearer and then further away. The person in the middle pointed to where they thought the sound was coming from. I thought it would be easy but it wasn't. When we'd been clapping we'd been clapping on the level of the ears so we clapped above and below the level of the ears. We found out that most people got the right direction but the wrong height."

Fig. 4.20

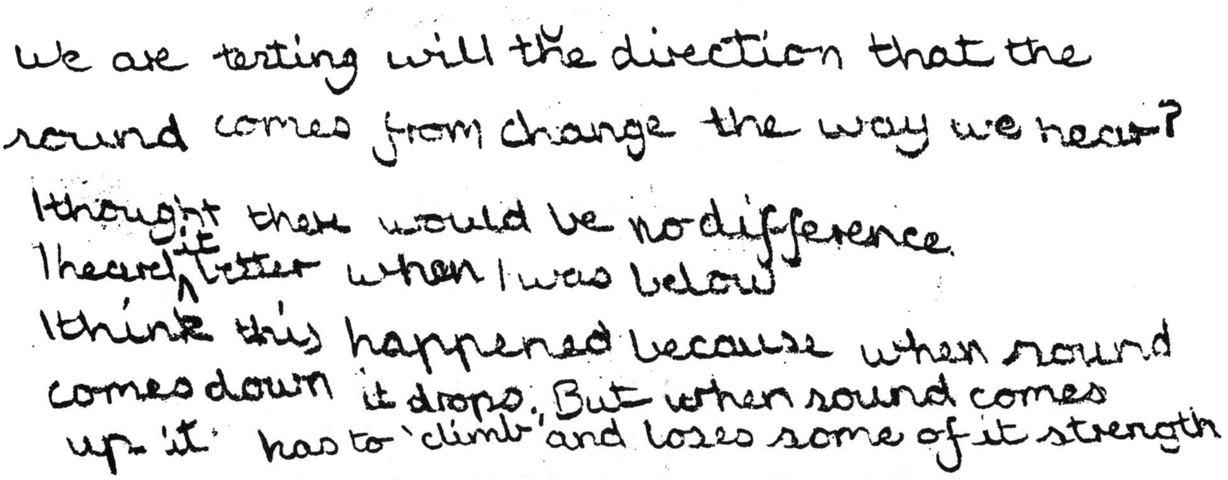

We are testing will the direction that the
sound comes from change the way we hear?
I thought there would be no difference.
I heard it better when I was below
I think this happened because when sound
comes down it drops. But when sound comes
up 'it' has to 'climb' and loses some of it strength

(Age 9 years)

> *"We are testing will the direction that the sound comes from change the way we hear? I thought there would be no difference. I heard it better when I was below. I think this happened because when sound comes down it drops. But when sound comes up it has to 'climb' and loses some of its strength."*

The explanation suggested in Fig. 4.20 could indicate that the child thought of the sound as something which would behave like an object with mass, like a ball or a person. The notion of sound transmission is implicit in this explanation.

Investigations concerning the transmission of sound were not suggested by any children in the two younger age groups. The string telephone was used as the context for one investigation attempting to block transmission (Fig. 4.21).

Fig. 4.21

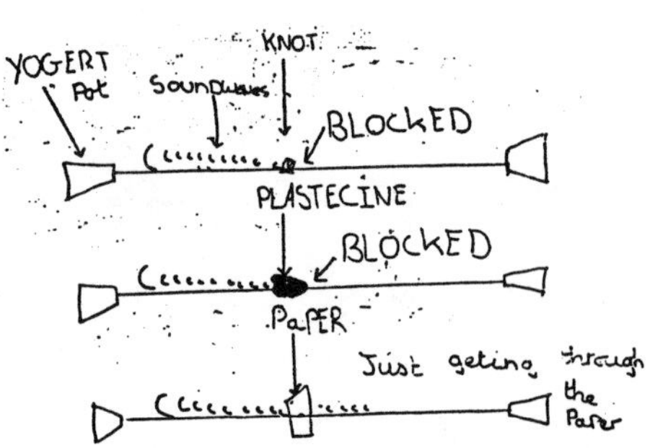

we are testing to see if a knot in a string telephone would stop the message From geting to the outher person
we put a knot in the string telephone and it afected are hearing as we guest it would. because the soundwaves can not get through the knot. Then we put some plastecine on. We thought the plastecine would block it off because the soundwaves could not get throught the plastecine. Then we guest that the paper would not block the soundwaves off we put the paper on and it only Just afected our hearing

(Age 9 years)

"We are testing to see if a knot in a string telephone would stop the message from getting to the other person. We put a knot in the string telephone and it affected our hearing as we guessed it would because the sound waves can not get through the knot. Then we put some plasticine on. We thought the plasticine would block it off because the sound waves could not get through the plasticine. Then we guessed that the paper would not block the sound waves off. We put the paper on and it only just affected our hearing."

Other children considered the passage of sound through water and through glass. During a class visit to the local swimming baths a group of children were able to test their ideas about hearing under water, and compare hearing through air and water. The children concluded that, to their surprise, they could hear the sound over a greater distance under water than through the air (Fig. 4.22).

Fig. 4.22

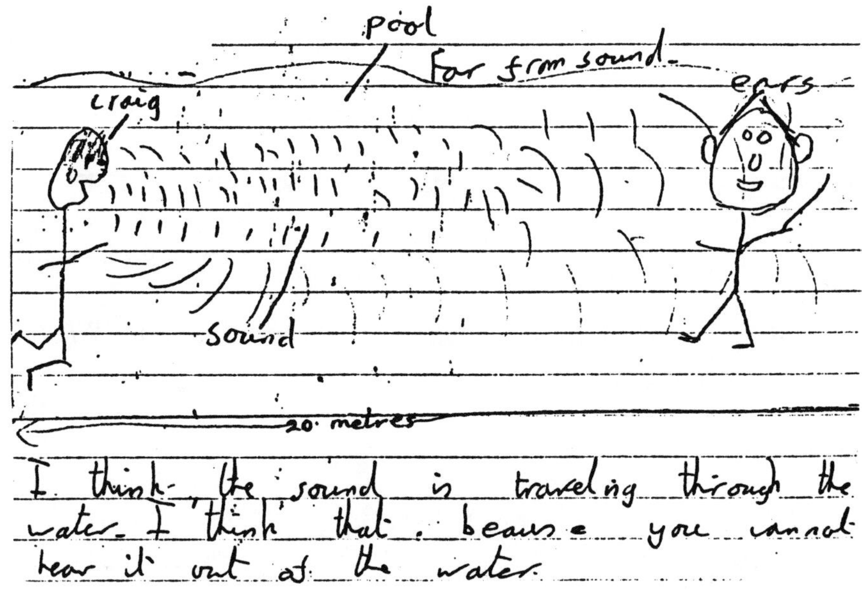

(Age 9 years)

"I think the sound is travelling through the water. I think that because you cannot hear it out of the water."

These investigations which produced results which surprised the children were challenging the notions which the children had about sound. The experiments which confirmed children's beliefs were also valuable in the way they encouraged the children to develop their ideas of sound through pictorial representation. The role of the teacher in encouraging children to interpret and explain their findings was very important and required teachers to develop their questioning skills in a manner complementary to that required for successful elicitation of ideas.

ii. Encouraging children to develop specific definitions for words they use

The upper junior children used a large number of words specific to sound: 'echo', 'vibration', 'interference', 'sound waves' and children were asked to define their meanings for these words. The range of definitions for 'echo' was wider than that given by the lower juniors (Figs. 4.23 to 4.26) but less reference was made to sound bouncing back and far more to sounds repeating themselves in empty spaces.

Fig. 4.23

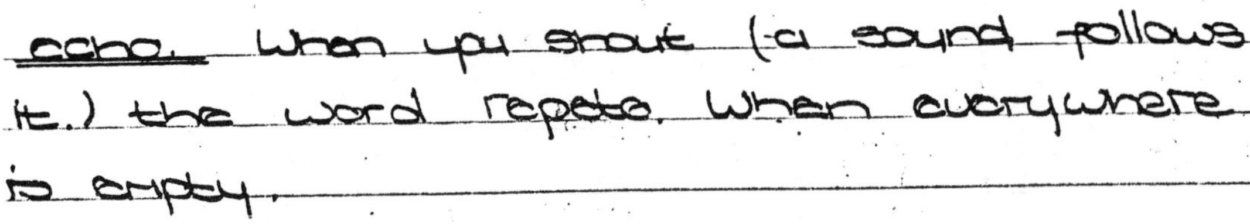

echo. When you shout (a sound follows it.) the word repeto. When everywhere is empty.

(Age 10 years)

"echo When you shout (a sound follows it) the word repeats. When everywhere is empty."

Fig. 4.24

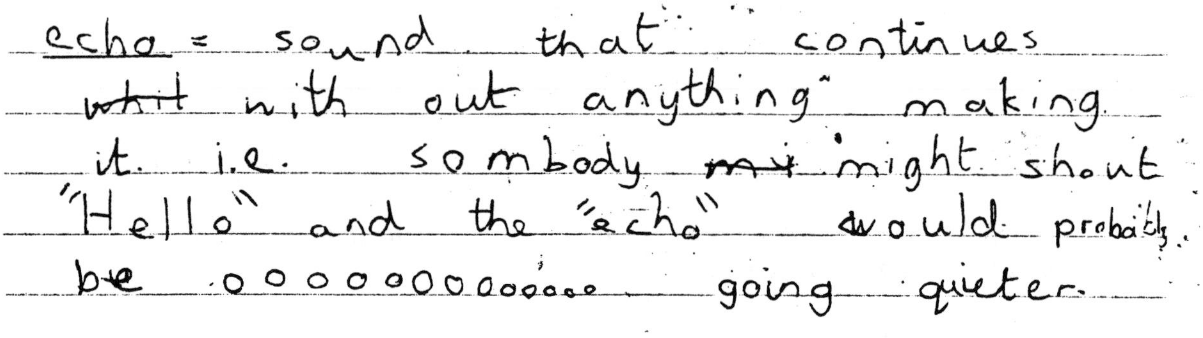

echo = sound that continues ~~whit~~ with out anything making it. i.e. sombody ~~mi~~ might shout "Hello" and the "echo" would probably be ooooooooooooo going quieter.

(Age 9 years)

"echo = sound that continues without anything making it i.e. somebody might shout "Hello" and the "echo" would probably be ooooooooooooo going quieter."

Fig. 4.25

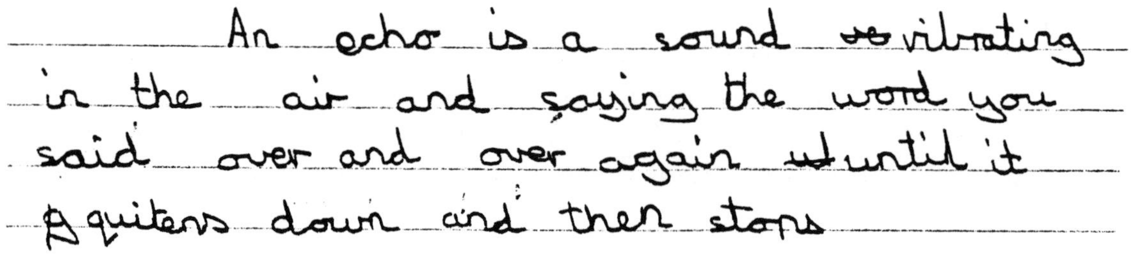

An echo is a sound vibrating in the air and saying the word you said over and over again until it quitens down and then stops

(Age 10 years)

"An echo is a sound vibrating in the air and saying the word you said over and over again until it quietens down and then stops."

Fig. 4.26

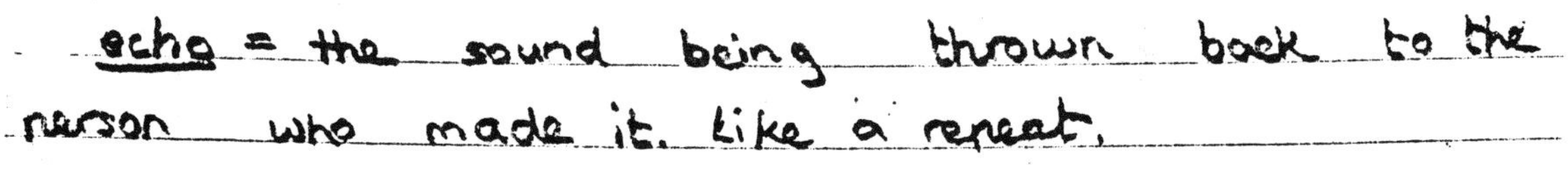

echo = the sound being thrown back to the person who made it. Like a repeat.

(Age 9 years)

"echo = the sound being thrown back to the person who made it. Like a repeat."

Definitions of vibration were far more closely linked to sound than were the lower junior definitions. Rather than emphasising the movement of the object, the term was commonly thought to be synonymous with sound, or involved with sound carrying (Fig. 4.27).

Fig. 4.27

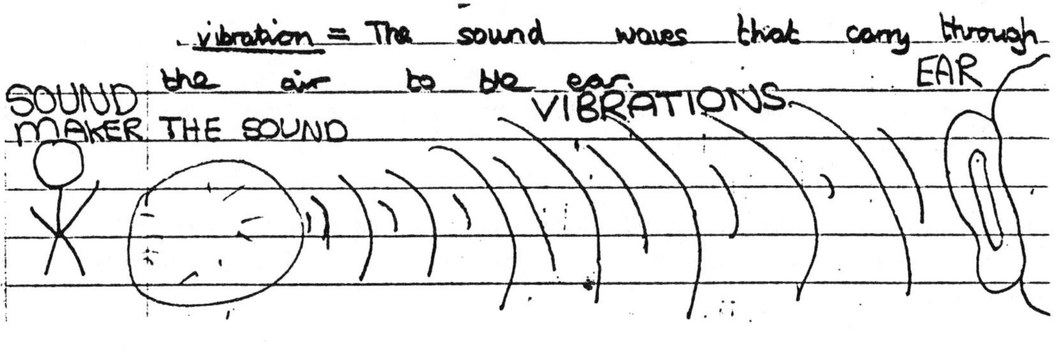

(Age 9 years)

"vibration = The sound waves that carry through the air to the ear."

Some definitions were suggestive of reference having been made to some secondary source (Fig. 4.28).

Fig. 4.28

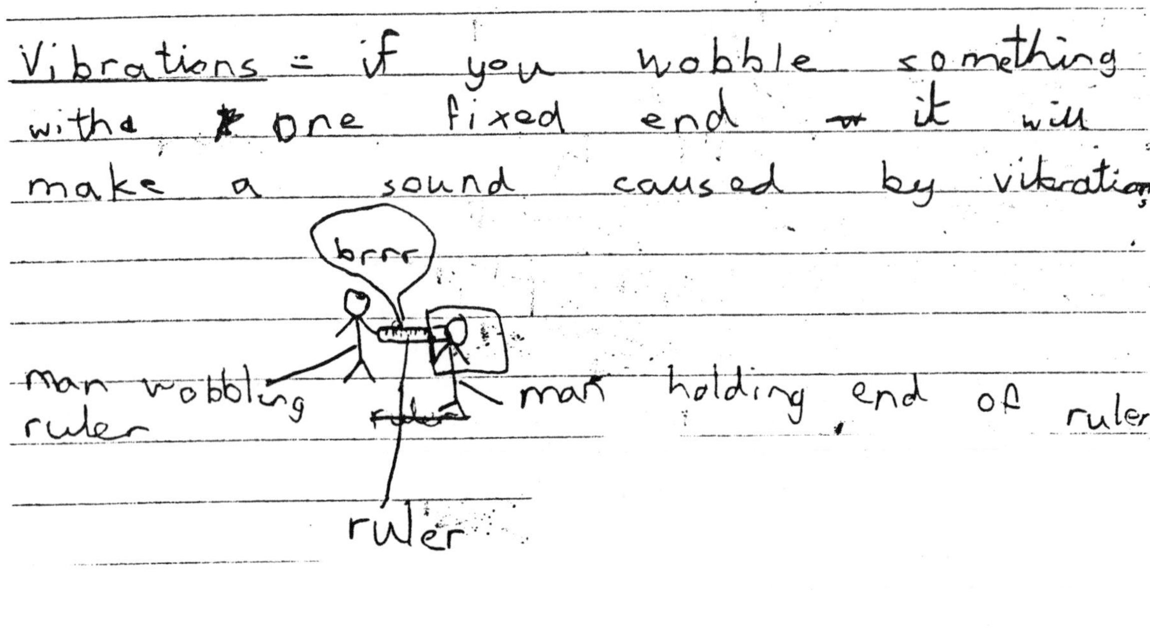

(Age 9 years)

"<u>*Vibrations*</u> *= if you wobble something with one fixed end it will make a sound caused by vibration."*

These definitions were produced by children as a written exercise prior to discussion so they were not based upon activities which could haven provided a common context for definitions.

iii. Encouraging children to generalize from one specific context

The upper junior children seemed very confident in exchanging contexts in which they had experienced the effects of interference, differences in volume and other variables upon hearing.

One teacher decided to see how far the class could generalize from some of the experiences which they had had during the Exploration. The children were posed the problem of trying to hear their partner's heart beat using some or all of the following equipment: yogurt pots, string, funnels, tube, scissors. The children tackled this task in a variety of ways, some making a string telephone, some using funnels connected with string and some using a funnel or a funnel and tube. This activity provoked lively discussion between the children and gave the teacher further insights into the children's understanding of sound, particularly of transmission through string.

Summary

1. Teachers used a question similar to "How can you affect the way you hear?" to introduce the Intervention to their classes.

2. The investigations chosen by children of different ages tended to be different. Younger children explored the reception of sound, and means of altering it. Many older children were interested in the effects of distance on sound reception, and in the effects of blocks at the point of production or of reception. Some upper juniors were interested in exploring sound transmission through different media.

3. Children were beginning to use some vocabulary particularly associated with sound in the lower juniors. The words 'echo' and 'vibrate' were often used to mean 'repeat' and 'wobble' respectively. Some confusion of 'vibration' with 'evaporation' was evident, and the converse was found during the topic 'Evaporation and Condensation'.

4. Children were able to generalize their classroom experiences of sound to everyday contexts.

5. The teacher's role during the Intervention was to encourage the use of science processes during children's investigations. With children who were more proficient in carrying out investigations the teacher was important in encouraging them to interpret their findings.

6. Investigations provided a helpful context in which children could develop a pictorial representation of sound.

5. THE EFFECTS OF INTERVENTION

Following the period of classroom Intervention work, those children who had been interviewed previously were seen again for discussion about their classroom activities and their ideas associated with sound. Since the Intervention involved children investigating what affected their hearing, the post-Intervention interviews had as their focus the questions associated with everyday sounds and children were asked at length about their investigations. The interviews took place during June 1988.

The questions relating to hearing everyday sounds did not address the production of sound directly but were more concerned with transmission and reception. However, response categories relating to sound production have been included where relevant, as some children offered incidental comments pertinent to sound production. The same categories as had been used in the Exploration phase were used again.

This chapter is composed of three sections: firstly, there is a comparison of the responses of the sample of children to interview questions about everyday sounds, pre- and post-Intervention. Secondly, there is an examination of changes in individual children's ideas as a result of the Intervention. Thirdly, some reaction of teachers to the project are discussed.

There was some decline in the interview sample size due to pupil absence. The sample from which pre- and post-Intervention interviews were completed contains:

12 infants (6 girls, 6 boys)
15 lower juniors (7 girls, 8 boys)
17 upper juniors (7 girls, 8 boys)
44 children (20 girls, 24 boys)

5.1 Changes in Children's Ideas

Children responded to the following questions with reference to the investigations which they had performed:

a. How can you hear the sounds around?

b. What makes a difference to how well you can hear?

Children's responses were frequently complex, incorporating more than one idea so the categories in Tables 5.1 and 5.2 are not mutually exclusive.

a. How can you hear the sounds around?

Table 5.1 shows a comparison of children's ideas about how sounds around could be heard, pre- and post-Intervention.

Table 5.1 A comparison of children's ideas before and after Intervention about how sounds around could be heard (Percentages)

	Pre-Intervention			Post-Intervention		
	Inf n = 10	LJ n = 14	UJ n = 12	Inf n = 10	LJ n = 14	UJ n = 12
Through the listener's ears	40 (4)	57 (8)	17 (2)	20 (2)	79 (11)	75 (9)
Don't know	20 (2)	14 (2)	33 (4)	-	-	-
Through the air	-	7 (1)	17 (2)	-	21 (3)	58 (7)
By active listening (no further explanation)	20 (2)	7 (1)	-	-	7 (1)	-
Due to the proximity of the sound to the listener	20 (2)	-	-	90 (9)	14 (2)	-
Due to the volume of the sound	-	7 (1)	8 (1)	-	-	-
Through thin barriers	10 (1)	7 (1)	8 (1)	-	7 (1)	-
Through holes in windows/ doors	-	7 (1)	-	10 (1)	21 (3)	8 (1)
No response	-	7 (1)	8 (1)	-	-	-
Other	-	-	8 (1)	-	-	8 (1)

Pre-Intervention a number of children had responded that they did not know how they heard the sounds around. No children gave this answer post-Intervention, and the increase in the number of children who were able to suggest an explanation for hearing was significant for the lower juniors ($p < 0.05$) and upper juniors ($p < 0.01$). This difference suggests that the children either had more ideas about sound as a result of the Intervention, or that they were more confident in expressing them.

There was a very large increase in the number of infants explaining hearing in terms of proximity to the sound source ($p < 0.01$).

There was an increase in the number of upper juniors mentioning ears as a reason for hearing ($p < 0.02$). As these categories are not mutually exclusive it is likely that this response was a part of a more complex answer mentioning sound transmissions (transcript 5.1) rather than a total response.

Transcript 5.1 Adrian, age 10

Q How did the sounds get to you?

A Through the air in sound waves and into the ear.

Q Can you tell me about sound waves?

A Well, sound travels through the air in waves - a bit like the sea at the edge.

Q How does it work?

A Sound can't go through so it has to go round or over. We can hear through glass because it's thin and there's little holes in wood so the sound creeps through the doors.

The range of responses given by upper juniors post-Intervention was more restricted than pre-Intervention, and significantly more upper juniors referred to sound travelling through air ($p < 0.05$). However, the example given in transcript 5.1 suggests that even when air is mentioned it is not necessarily regarded as a medium through which sound travels. The use of the term 'sound wave' in the same example could also be indicative of formally acquired knowledge.

The number of children who suggested that sound travelled through thin barriers or through holes in windows and doors did not change appreciably as a result of the Intervention. It was an idea held by a small number of children within each age group, with the largest number of responses at the lower junior age.

The number of children mentioning active listening as the mechanism by which sounds were heard was not significantly different before and after Intervention.

b. What makes a difference to how well you can hear?

The data in Table 5.2 show a general trend towards more children considering a greater number of conditions as relevant to hearing, post-Intervention.

Table 5.2 A comparison of children's ideas about conditions which affect hearing, before and after Intervention (Percentages)

	Pre-Intervention			Post-Intervention		
	Inf	LJ	UJ	Inf	LJ	UJ
	n = 12	n = 15	n = 16	n = 12	n = 15	n = 16
Distance	50 (6)	40 (6)	44 (7)	75 (9)	53 (8)	56 (9)
Volume	42 (5)	53 (8)	25 (4)	17 (2)	53 (8)	31 (5)
Background noise	42 (5)	13 (2)	38 (6)	8 (1)	33 (5)	38 (6)
Obstacles	8 (1)	-	13 (2)	17 (2)	40 (6)	38 (6)
Listening	-	-	19 (3)	8 (1)	13 (2)	6 (1)
Pitch	-	-	-	-	-	13 (2)
Concentration on work	-	-	-	-	7 (1)	-
Looking			6 (1)	-	-	-
Nothing	17 (2)	13 (2)	-	8 (1)	-	-
Don't know	8 (1)	7 (1)	6 (1)	8 (1)	7 (1)	-

The trends in the table, while generally towards children mentioning more conditions, show a significant increase in reference to one condition only: post-Intervention, 40% of lower juniors said that obstacles in the path of the sound would affect the way the sound was heard (0% pre-Intervention, $p < 0.02$). The majority of references to obstructions of sound were related to investigations with ear muffs. Some children had also extended their understanding to include types of obstacle which would block the path of the sound to the ear (transcript 5.2)

Transcript 5.2 Michael, age 10

Q *What did you do?*

A *One person played one sound all the time. We listened to it without anything muffling. Then we tried ear muffs. They were a good sound blocker, and the sound was not as good as without......*

......When you're in another room and someone speaks to you you can't hear as only some of the sound gets through; the rest hits the door or wall and rebounds into the room and goes back to the speaker.

This increase in frequency of upper and lower juniors reporting obstacles as a condition which would affect hearing ($p < 0.02$) suggests that more children were starting to consider sound transmission.

Table 5.3 shows the number of conditions mentioned by each child and whether these conditions were associated with sound production, transmission or reception. There was a small trend towards older children mentioning more conditions, and the responses from the two junior age groups showed a trend towards mentioning a greater number of conditions post-Intervention. The most noticeable differences between the pre- and post-Intervention interview responses were concerned with the number of conditions relating to sound transmission. More transmission conditions were mentioned post-Intervention, and the increase was significant with respect to the upper junior sample ($p<0.02$).

Table 5.3 A comparison of the number of conditions mentioned by individual children pre- and post-Intervention (Percentages)

		Pre-Intervention			Post-Intervention		
		Inf n = 12	LJ n = 15	UJ n = 16	Inf n = 12	LJ n = 15	UJ n = 16
Total number of	0	17	27	-	8	-	19
conditions mentioned		(2)	(4)		(1)		(3)
	1	33	33	50	50	33	38
		(4)	(5)	(8)	(6)	(5)	(6)
	2	17	27	38	25	33	13
		(2)	(4)	(6)	(3)	(5)	(2)
	3	25	7	6	8	20	19
		(3)	(1)	(1)	(1)	(3)	(3)
	4	-	-	-	-	7	13
						(1)	(2)
	Don't know	8	7	6	8	7	-
		(1)	(1)	(1)	(1)	(1)	
Mean number of conditions mentioned		1	1.1	1.4	1.3	1.9	1.7
Number of conditions	0	50	40	69	75	40	63
relating to sound		(6)	(6)	(11)	(9)	(6)	(10)
production	1	42	53	25	17	53	31
		(5)	(8)	(4)	(2)	(8)	(5)
	2	-	-	-	-	-	6
							(1)
	Don't know	8	7	6	8	7	6
		(1)	(1)	(1)	(1)	(1)	(1)
Number of conditions	0	33	53	50	17	20	19
relating to sound		(4)	(8)	(8)	(2)	(3)	(3)
transmission	1	58	40	31	58	53	75
		(7)	(6)	(5)	(7)	(8)	(12)
	2	-	-	13	17	20	6
				(2)	(2)	(3)	(1)
	Don't know	8	7	6	8	7	-
		(1)	(1)	(1)	(1)	(1)	
Number of conditions	0	50	80	38	75	47	63
relating to sound		(6)	(12)	(6)	(9)	(7)	(10)
reception	1	42	13	50	17	40	31
		(5)	(2)	(8)	(2)	(6)	(5)
	2	-	-	6	-	7	6
				(1)		(1)	(1)
	Don't know	8	7	6	8	7	6
		(1)	(1)	(1)	(1)	(1)	(1)

Sound Production

The notion of sound being produced was not one which was addressed by the Intervention or by the post-Intervention interviewing and children's incidental comments showed very little difference in ideas about sound production.

Sound Transmission

The ideas about sound transmission were mentioned by significantly more children after the Intervention than before (lower junior, $p < 0.05$, upper junior, $p < 0.001$). Additionally, more upper juniors referred to sound travelling through a medium ($p < 0.01$). The figures in Table 5.4 suggest that the Intervention activities in which the children were involved encouraged them to think about how the sound moved from its source to where it was heard.

Table 5.4 A comparison of the incidence of reference to sound transmission before and after Intervention (Percentages)

	Pre-Intervention			**Post-Intervention**		
	Inf n = 12	**LJ n = 15**	**UJ n = 17**	**Inf n = 12**	**LJ n = 15**	**UJ n = 17**
No mention of sound travelling	100 (12)	67 (10)	71 (12)	83 (10)	27 (4)	12 (2)
Mention of sound travelling, but no mention of medium	-	27 (4)	12 (2)	17 (2)	53 (8)	35 (6)
*Mention of sound travelling through air	-	7 (1)	18 (3)	-	20 (3)	53 (9)
*Mention of sound travelling through a medium	-	-	-	-	-	18 (3)

*not mutually exclusive

The differences between pre- and post-Intervention interviews seem to increase with age, and it is likely that sound travelling, an imperceptible phenomenon, is more accessible to children who have had a broader range of experiences concerning sound, and where development has reached a stage where more abstract thought is possible.

The idea that sound travels through a medium, often air, was mentioned more frequently post-Intervention, but it was not always obvious that children considered air to be a medium, as can be seen in transcript 5.3. The child who was being interviewed seemed to feel that the air would not obstruct the sound in the same way as the water.

Transcript 5.3 Craig, age 10

Q *Did you expect to hear under water?*

R *No, we were surprised.*

Q *Did you think it would be better under water?*

R *Yes, it was quieter. but if there was no noise the same distance in the air I think it would be better in air 'cos sound travels faster in air than water. It has to travel through something in water, it doesn't in air.*

Q *Can you explain what 'travelled' means?*

R *The air which comes out of your mouth pushes along until it gets to the people. A man came a long time ago and told us about dolphins and whales hearing.*

Q *Have you changed your ideas at all?*

R *Yes. You can hear well under water, but if there was not noise in the air you'd hear better in air.*

Sound Reception

Table 5.5 shows there were substantial differences between pre- and post-Intervention interview responses concerning the reception of sound.

Table 5.5 A comparison of the incidence of reference to sound reception before and after Intervention (Percentages)

	Pre-Intervention			Post-Intervention		
	Inf n = 12	**LJ** n = 15	**UJ** n = 17	**Inf** n = 12	**LJ** n = 15	**UJ** n = 17
No mention of a sound receptor	75 (9)	40 (6)	77 (13)	67 (8)	13 (2)	29 (5)
Mention of the ear as a receptor	25 (3)	53 (8)	24 (4)	33 (4)	80 (12)	65 (11)
Mention of vibrations being set up in the receptor	-	7 (1)	-	-	7 (1)	6 (1)

After the Intervention, more children of every age mentioned the ear and the differences from pre-Intervention were more pronounced with the older children (upper juniors $p < 0.01$). The Intervention activities might have encouraged children, through focusing on the concept of sound, to consider the parts of their bodies which were responsible for receiving sound. Very little mention was made of vibrations being set up in the receptor either before or after the Intervention.

Descriptions of sound

Children of junior age were far more likely to use a verbal representation for sound travelling after the Intervention than before (lower juniors $p < 0.1$, upper juniors $p < 0.001$).

Table 5.6 A comparison of the descriptions used by children about sound before and after Intervention (Percentages)

	Pre-Intervention			Post-Intervention		
	Inf n = 12	LJ n = 15	UJ n = 17	Inf n = 12	LJ n = 15	UJ n = 17
No description used	100 (12)	67 (10)	77 (13)	83 (10)	33 (5)	12 (2)
Sound described as 'sound' travelling	-	27 (4)	24 (4)	17 (2)	27 (4)	41 (7)
Sound described as 'words'	-	7 (1)	-	-	7 (1)	-
Sound described as 'sound waves'	-	-	-	-	20 (3)	41 (7)
Sound described as 'vibrations'	-	-	-	-	-	6 (1)
Sound described as 'echoes'	-	-	-	-	13 (2)	-

This increase in the use of a descriptor might indicate that children had needed to construct a representation when exploring ideas about sound, particularly where their investigations led them to develop notions of sound transmission which had not been present before.

5.2 Changes in Individual Children

It was possible to identify the changes in the type of response given by individual children, pre- and post-Intervention, using the broad groupings of ideas about sound production, transmission and reception which had been identified during the Exploration phase.

Sound production

1. No action There is no mention that an action/movement/input of energy is necessary in order for sound to be produced.

2. Action	There is a clear indication that an action/movement/input association of energy is necessary in order for sound to be produced.
3. Vibration	There is considered to be an association between vibrations and sound production.

Sound transmission

1. No travel	There is no mention of sound travelling.
2. No medium	Sound is mentioned as travelling, but there is no mention of a medium through which it travels.
3. Travel through air	Sound is described as travelling through air.

Sound reception

1. No receptor	There is no mention of a sound receptor.
2. Ear as receptor	The ear is mentioned as the sound receptor.
3. Receptor vibration	Vibrations are thought to be set up within the receptor.

Sound Production

Table 5.7 The direction of changes in the ideas of individual children about sound production, pre- and post-Intervention.

pre- → post-Intervention	Infant n = 12	Lower Junior n = 15	Upper Junior n = 17
1 → 1	9		3
2 → 1	3	5	4
2 → 2		7	9
2 → 3		3	1
3 → 3			

From Table 5.7 it can be seen that the age group in which there was most change (53%) is the lower juniors.

Some of these changes were towards more scientifically acceptable responses (20%) while others were not obviously in that direction. Once again, it must be stressed that the Intervention activities and interview questions did not focus directly on sound production, and any mention was incidental. From these limited data it appears that 'action association', mention of the need for energy/action/movement in order for sound to be produced, is the idea from which all of the change occurs. This suggests that children need to understand that sound production requires an action before they can understand about vibrations.

Sound Transmission

Table 5.8 The direction of changes in the ideas of individual children about sound transmission, pre- and post- Intervention.

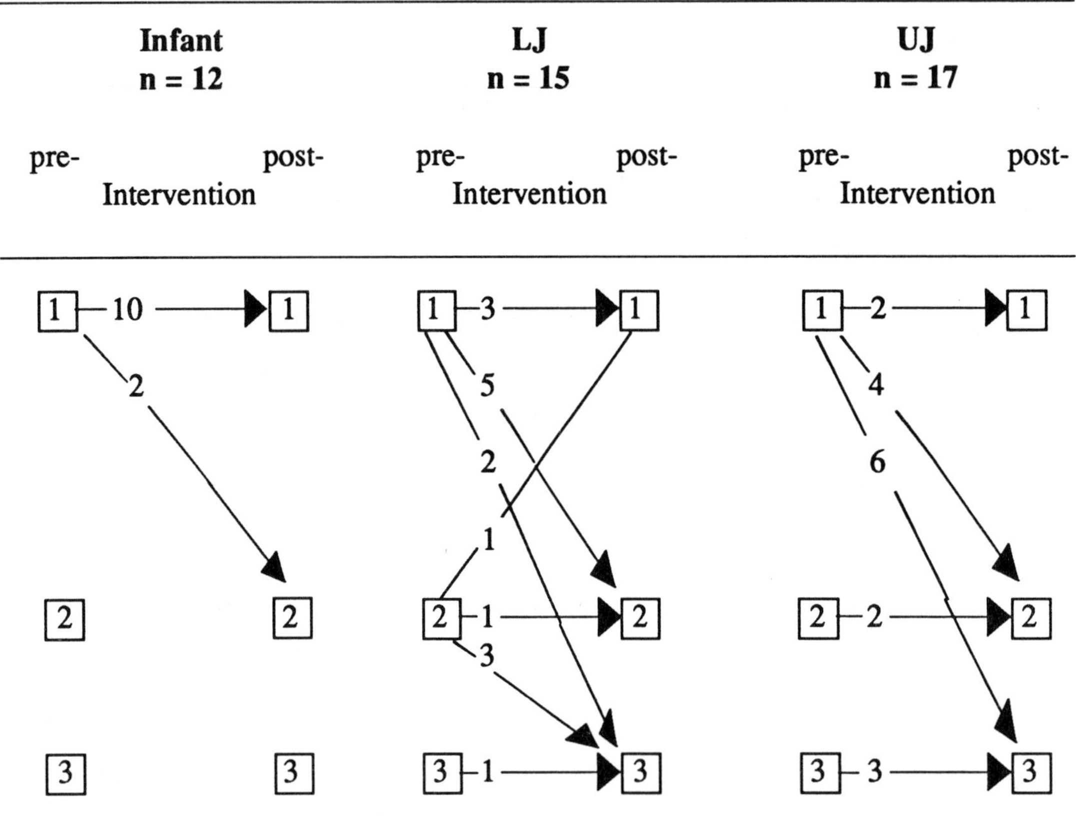

From table 5.8 it can be seen that there was very little change (17%) in the ideas which infants held about sound transmission. Both of the junior groups showed a high percentage of change from a 'no travel' response to a 'no medium' or 'travel through air' 3 response. Only one lower junior changed from a 'no medium' to a 'no travel' response. A quarter of the junior sample expressed ideas which were of 'no travel' pre-Intervention and post-Intervention were of 'travel through air'. This apparent omission of 'no medium' responses is ambiguous. It could suggest that the 'no medium' response, that sound travels but without any mention of a medium (or in the absence of any obstacles) need not be a precursory idea to the notion that sound travels through air. Alternatively, the 'travel through air' response might indicate learning from a secondary source, including peers (transcript 5.4), and may in some cases be indistinguishable from a 'no medium' response except for the label 'air'.

Transcript 5.4 Timothy, age 10

Q *How do you think we hear?*

R *By the sound which is coming into the ear.*

Q *Can you explain that to me?*

R *The person claps and it travels to the ear. It goes to the ear drum and into the brain.*

Q *Can you explain 'it travels'?*

R *It goes by air to the ear.*

Q *Can you tell me a bit more about that?*

R *I don't really know.*

Q *Can you explain travels?*

R *It doesn't mean like I travel in a car it means - oh I don't know - but it's not travelling in a car or like me on a bike - it's something else.*

Each of the children who suggested that sound travelled through a medium post-Intervention but not pre-Intervention had been engaged in an investigation which could have encouraged the children to consider sound travelling:

. the effect of distance from the sound source on hearing (three children)

. blocking the reception of sound (two children)

. the effect of the orientation of the listener to the sound source

. the effect of interference

. whether sound could be heard under water (transcript 5.5)

Transcript 5.5 is taken from an interview with a child who was trying to reconcile the findings of his investigation with his previous ideas.

Transcript 5.5 Daniel, age 10

Q *Have you learned anything from your investigation?*

R *I didn't think sound travels as far in water as it did.*

Q *Did it travel as far in air?*

R *About the same.*

Q *Why did you expect it not to travel as far in water?*

R *Probably because water is like a block - it still gets through but not as clear.*

The post-Intervention interview data does not allow even tentative conclusions to be drawn about the context specificity of these responses since the interviews were mainly about children's Intervention activities and it would be necessary to re-interview the children about the Elicitation activities to estblish whether children could generalise the effects of the Intervention to the different contexts of the interview. Of the two infant children who were re-interviewed about the Elicitation activities and who changed their response there was little evidence of a generalization in understanding. Transcript 5.6 records a child's ideas following an investigation on the effect of distance on hearing.

Transcript 5.6 Neal, age 6

Questions about child's investigation on the effect of distance on hearing.

Q *What did you find out?*

R *The sound got lower when I went further away. It was still there even when I couldn't hear it.*

Q *What happens to the sound when you can't hear it?*

R *The sound will bounce away. At the 'Backs' there's a wall and the sound comes back because it can't get away from the wall. The sound bounces off the walls - they're too strong for the sound to get away.*

Q *So what happens if you go too far away to hear the sound?*

R *The sound must have bounced back.*

Q *How does the rubber band make a sound?*

R *The air must make the sound as well because the air's in the middle and it must make the air go so fast that it makes a noise as well. The rubber band pushes the air and it makes a sound. If you swing anything like a belt or a bit in front of you it makes a noise because the air's being pushed.*

Q *How do you think that sound gets to you so you can hear it?*

R *I don't know.*

Sound Reception

Table 5.9Changes in children's ideas about sound reception, before and after Intervention.

Infant n = 12	LJ n = 15	UJ n = 17
pre- post- Intervention	pre- post- Intervention	pre- post- Intervention
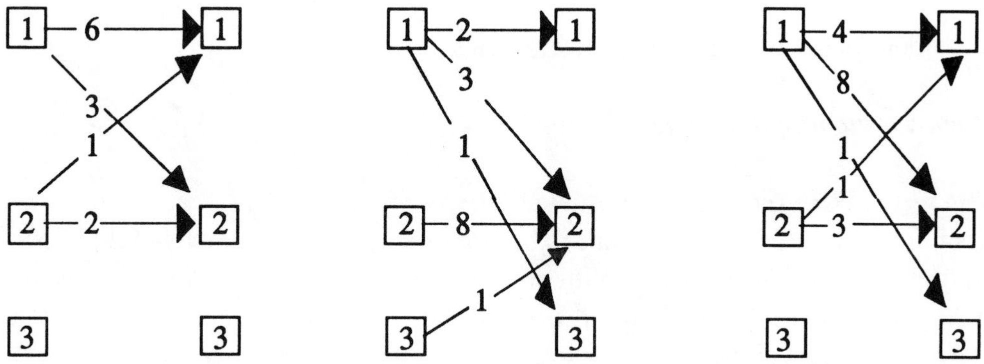 1 –6→ 1 3 1 2 –2→ 2 3 3	1 –2→ 1 3 1 2 –8→ 2 1 3 3	1 –4→ 1 8 1 1 2 –3→ 2 3 3

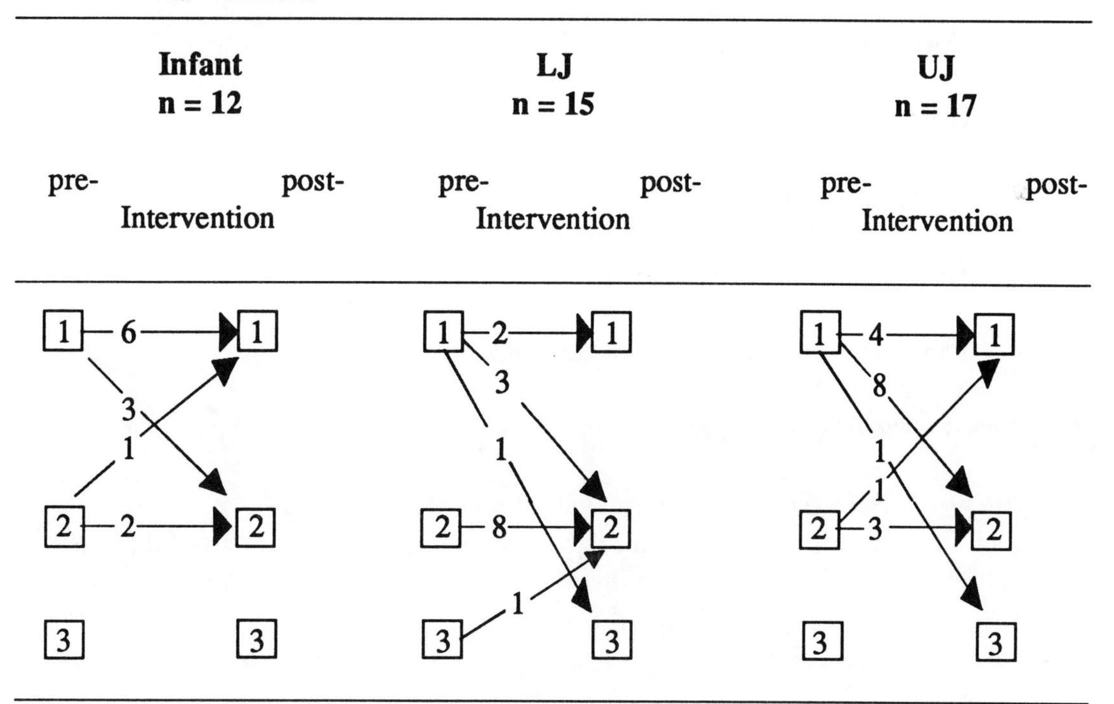

Table 5.9 shows that the proportion of children whose ideas about sound reception changed was approximately one half in each age group. Most changes occurred amongst the upper juniors where 53% of children changed the type of response they gave. Post-Intervention, 42% of infants and 75% of juniors mentioned the ear during interviews about 'sound'. Very few children mentioned the need for vibrations to be set up in the ear in order for sound to be heard. The lower junior who had given that response before Intervention did not repeat it post-Intervention; his investigation was

concerned with blocking the transmission of sound. The two children who stated this idea for the first time post-Intervention had carried out investigations into:

. blocking the reception of sound

. the effect of distance, and using one or two ears, on hearing

Both of these investigations encouraged children to focus on the reception of sound. It is not possible to perceive sound waves hitting the ear drum and making it vibrate so the Intervention might have meant that the children were motivated to seek out the information for themselves in books.

5.3 Teacher's Reactions to Project Involvement

Changes in attitudes and classroom practice in teachers need to be considered in the context of the whole research programme, rather than the Intervention alone.

The main areas in which teachers reported change, or the Project team perceived it, were:

1. Open questioning techniques
2. Use of annotated diagrams
3. Role in class discussion
4. Familiarity with science process skills
5. Identification of Intervention strategies

1. Open questioning technique

Teachers were asked to talk to individual children to find out their ideas. The teachers initially found it very difficult to phrase open questions which would allow the children to express their ideas. With practice this became easier, and as a result the teachers were more relaxed about the questioning and this enabled them to listen more carefully to the children's responses. The teachers' ability to build upon these answers in the developing dialogue and to clarify the children's meaning also improved. Some teachers began to find this technique so valuable to their science work that they took every opportunity to question children, including informal conversations during breaks in the school day. Other teachers began to incorporate it into their work in other curricular areas, using it as a way of increasing children's participation in language and art work (transcript 5.7).

Transcript 5.7

"As this project has progressed I have become increasingly aware of the effect it has had on me as a teacher. Initially, I had to think very carefully when talking to the children about the questions I was asking. This made the task very difficult, as years of different approaches had to be avoided i.e. the teaching situation. However, as time has passed it has become part of my technique and in my opinion an invaluable part.....

"By the use of careful questioning you are able to extend these ideas without feeding them information they are not ready for. In science, in particular, it is difficult to know how deeply to go into a topic, but by using this approach the children give you the boundaries. This must be a far more relevant approach.....

"Certain areas of the curriculum will have to be taught but by adapting this technique a much more flexible and interesting approach can be made to many of the subjects. It is certainly going to have an effect on my methods.

" This week I am taking an artefact into school to look at with the children. Before this project I would have told the children all about it and expected them to sit and listen. Now, I shall use this method of questioning to find out the children's ideas and hopefully use their imagination too....."

2. *Use of annotated diagrams*

The Project team requested that teachers asked their classes to draw particular diagrams as part of the Project work. As the Project progressed, some teachers began to extend the number of diagrams which children were asked to draw during the Elicitation phase in order to give themselves a fuller understanding of what were their children's ideas. Other teachers extended the use of the technique into the Intervention phase, using drawings as a starting point for vocabulary work. Some attempt was also made to use the diagrams as an assessment tool, asking children to draw a diagram in response to the same question both before and after the Intervention.

3. *Role in class discussion*

Many of the teachers who were involved with the Project felt that class discussion was something with which they were familiar. However, these discussions were often very teacher-directed and tended to be question and answer sessions where the answers were either right or wrong. By posing a more open question, teachers found that the children were able to contribute far more to discussions, and that they listened more to the contributions others were making. The children began to spark off ideas in each other, and the discussions became far more self-sustaining. The role of the teacher became more that of a chairperson, ensuring that more reticent children could participate, and asking a few, well-chosen questions to demarcate the area to be covered during the discussion. Class discussion therefore became another opportunity for the teacher to learn about the class by listening to their contributions.

4. *Familiarity with science process skills*

Teachers themselves become more conversant with the process approach to science. This increased their confidence in intervening when children were planning and carrying out investigations and enabled them to ensure fair testing, reliable data collection and valid interpretations.

5. *Identification of intervention strategies*

During Phase 1 of the Primary SPACE Project the teachers were at times unsure of what would constitute a 'vocabulary' activity or a discussion to encourage generalizations. With the experience of the 'Growth' Intervention (in which all 'Sound' teachers had participated) the teachers became far more confident in identifying different components within one complex classroom activity. 'Generalizing' was identified as being a natural part of many discussions and 'vocabulary' was often closely linked with it.

Summary

1. The post-Intervention interviews centred upon the Intervention activities with which the children had been involved in the classroom, and the questions which were addressed were:

   i How do the sounds around reach you so you can hear them?

   ii What makes a difference to how well you can hear?

2. There were significant changes in the ideas which children expressed about sound. These changes were largest in connection with sound transmission. The age group in which changes were most pronounced was the upper juniors.

   i More junior children mentioned that sound travelled (lower junior $p<0.05$, upper junior $p<0.001$).

   ii More upper junior children referred to sound travelling through a medium ($p<0.01$)

   iii More upper junior children used descriptions to elaborate their discussions about sound travelling.

3. An examination of the changes within individual children showed that a large number of children had developed their ideas in a way which could help them to develop more scientific thinking. While many children's ideas remained unchanged there were very few children whose ideas were less explanatory than prior to the Intervention.

4. Through participating in the SPACE Project many teachers had developed skills in non-directive classroom techniques and some were becoming more analytical in their approach to teaching.

6. SUMMARY

This section contains a summary of the main findings of the research into the topic of 'Sound'. Some implications for science education are also considered.

The research phases placed great emphasis on classroom activities undertaken by pupils and teachers. This work enabled children to express their ideas to both teachers and researchers in structured and less formal ways, and then to work with confidence with their ideas.

The Exploration activities were each very different in the experiences which they provided, but each of them was profitable in terms of encouraging children to express their ideas concerning particular aspects of sound. The pupils and teachers were both more familiar with expressing ideas and encouraging their expression. This familiarity enabled teachers to feel confident about their class's ideas and to develop a picture of possible avenues which the children could profitably investigate during the Intervention phase in their classrooms.

The main ideas which children put forward during pre-Intervention elicitation, both class-based work and individual interview, were as follows:

1. 'Sound' and 'vibration' were not intuitively linked by young children but an association developed as children's experiences broadened. Whether sound caused vibrations or vibrations caused sound seemed to depend upon the context. Some children suggested that sound and vibration were the same.

2. The production of sound from an object was often attributed to the properties of the object or to an impact. Children suggested mechanisms for the generation of sound from a drum. These mechanisms often involved vibrations and the site of sound production was suggested to be either inside the drum or at the surface of the drum.

   Some models mentioned vibrations both inside the drum - at the drum surface, in explanations which were similar to those accepted in conventional science.

3. Sound transmission was not an idea which was expressed by many young children. Infant children said they heard sounds because of the volume, the proximity to the sound source or because of a characteristic of the listener. Where sound travel was mentioned there was a prevalent idea that sound needed an unobstructed path along which to travel. Some older children considered sound to

travel through the string on the string telephone, or through air. What was meant by 'air' was often unclear.

4. Sound reception was frequently associated with the ear. A funnel ear-trumpet was thought to affect hearing by either collecting or concentrating sound so that more reached the ear. A small number of junior children mentioned the ear drum and the brain in connection with hearing. Some of these children mentioned vibrations being set up in the ear drum.

5. A small number of children explained being able to hear in terms of a psychological model, 'active learning'. They considered that the important determinant of hearing was that the listener was attending to the sound source. A similar notion was held by children in relation to vision.

6. A wide range of representations was used in diagrams to portray sound. Some were idiosyncratic and parallel to the direction of sound travel, and others more like the accepted scientific notation, perpendicular to the direction of travel. A small number of children portrayed sound spreading out from the source. The idiosyncratic notations could have been influenced by the representations of sound shown in children's comics. During interview, very little reference was made to the form in which sound travelled.

7. Junior children made frequent use of words such as 'vibrate', 'echo', 'travel' and 'sound wave' in association with descriptions of ideas about sound. Both 'vibrate' and 'echo' were often used in a manner which implied a meaning of 'repeat'.

A preliminary inspection of the children's ideas led to the formulation of a question which would be an appropriate starting point for the classroom Intervention work. The question was worded similarly to "How can you affect the way you hear?"

The investigations chosen by children of different ages tended to be different. Younger children explored the reception of sound, and means of altering it. Many older children were interested in the effects of distance on sound reception, and in the effects of blocks at the point of production or of reception. Some upper juniors were interested in exploring sound transmission through different media.

The teachers were more familiar with their role as managers of learning during the Intervention because of their earlier experiences with the Project in the topic of 'Growth'. This role involved teachers in encouraging the use of science processes during children's investigations. Where children were more proficient in carrying out investigations the teacher was important in encouraging them to interpret their findings. Children were making careful observations and by asking them to explain their findings, the teachers were helping the children to challenge their ideas. Additionally, the investigations provided a helpful context in which children could develop a pictorial representation of sound. The depiction of ideas about sound, particularly sound transmission, also encouraged children to explore their ideas further. The drawings which accompanied investigations provided a useful discussion point for teachers and children to clarify the interpretation of observations and to lead to the suggestion of further investigations.

The post-Intervention interviews centred upon the Intervention activities with which the children had been involved in the classroom, and the questions which were addressed were:

i. How do the sounds around reach you so you can hear them?

ii. What makes a difference to how well you can hear?

There were significant changes in the ideas which children expressed about sound. These changes were largest in connection with sound transmission. The age group in which changes were most pronounced was the upper juniors.

i. More junior children mentioned that sound travelled (lower junior $p<0.05$, upper junior $p<0.001$).

ii. More upper junior children referred to sound travelling through a medium ($p<0.01$)

iii. More upper junior children used descriptions to elaborate their discussions about sound travelling.

An examination of the changes within individual children showed that a large number of children had developed their ideas in a way which could help them to develop more scientific thinking. While many children's ideas remained unchanged there were very few children whose ideas were less explanatory than prior to the Intervention.

Through participating in the SPACE Project many teachers had developed skills in non-directive classroom techniques and some were becoming more analytical in their approach to teaching. The Project had provided teachers with a structure around which to base classroom work and because of the support which regular meetings were able to provide, many teachers felt confident to incorporate new techniques into their teaching. Again, regular meetings enabled teachers to compare outcomes and to see common ground where it existed between different ages of children and different school backgrounds. The fostering of an open questioning atmosphere within classrooms was enjoyed by pupils and teachers and led to both parties listening to what children were saying.

The implications of the Primary SPACE Project research on 'Sound' go beyond informing on the prevalent ideas which children hold about sound. They suggest that the elicitation techniques developed by the Project are not only feasible for classroom use, but useful to teachers as additions to their repertoire of informal assessment techniques. The development of science process skills is also integral with the approach of children working from their own ideas to lead to conceptual development.

APPENDIX I

SPACE SCHOOL PERSONNEL

LANCASHIRE

County Adviser for Science: Mr P Garner

School	Head Teacher	Teachers
Moor Nook Primary	Mr G Robinson	Mrs M Harrison
		Mrs L McGuigan
Farington Primary	Mr P S Warren	Mr J Evans
		Mrs M Pearce
St Teresa's RC Primary	Mr L Rigby	Mrs A Hall
		Mrs R Morton
Frenchwood Primary	Miss E M Cowell	Mrs C Pickering
Clough Fold Primary	Mr G M Horne	Mrs J Looker
		Mrs C Murray
Longton Junior	Mr J R Doran	Miss R Hamm
		Mrs L Whitby

APPENDIX II

Exploration Criteria

Teachers were actively involved in the construction of the programme for the second round of topics. As the lead-in to the design of activities for the Exploration phase, the criteria for suitable classroom experiences were considered by each of the three groups, and were later amalgamated under the following headings:-

Familiarity and relevance to children's experiences

Clear relationship between Exploration activity and other experiences. Relevant and familiar material, based on children's everyday experiences, relating to their awareness of the world around.

Safety

The activity should be safe. (For example, avoid the opportunity for children to put things in their ears and nostrils; avoid drinking or smelling liquids; avoid sharp objects and glassware).

Durability

The activity should be sufficiently durable to last the full extent of the Exploration phase. Plant or animal material should be robust and hardy.

Interest

The activity should be enjoyable, motivating and appealing, capable of stimulating interest and of making an immediate impact on children.

Simple Resources

The activity should be capable of being constructed from junk or easily available everyday objects and should involve simple technical processes which can be easily set up and maintained by teachers.

Accessibility

Activities should involve concrete examples, something large enough to be observed or handled easily. If the point of the resource material is that it changes in some way, the change should be speedy enough to be perceptible. The quality or degree of change must be obvious, something which the children can notice readily.

Comprehensibility

The outcome of the activities should be comprehensible to teachers and capable of eliciting clearly targetted areas of thinking in children.

Extendibility

The activity should lend itself to further activity, thinking, progression or investigation (during the Intervention phase). This may imply that obviously multi-variable situations should be avoided.

Ethical

When living material is used, it should be handled with regard to ethical considerations.

Replicability

The activity should be capable of being repeated in a similar way in all classrooms, with the means for replication between schools for research purposes.

Pupil Self-Direction

The activity should make minimal demands on teacher oversight, which implies that it should be sufficiently self-evident to be pupil-directed.

Recordable Responses

The activity should give rise to recordable pupil responses e.g. a drawing or some other brief form of recording.

Unobtrusive

The activity should be sufficiently compact to be useable in busy crowded classrooms.

APPENDIX III

Teacher Guidelines for SPACE Exploration - Phase II (March 1988)

We intend to give the children a chance to think about the new topic area before trying to find out what their ideas are.

The exploration phase should be very low-key, simply putting a range of experiences before the children so they've got something around which to structure their thoughts. We are doing the exploring of their ideas at this stage. Every child should have the opportunity to observe, consider and record their ideas. Some children may wish to go further and they should be allowed to do so but fight your instincts to teach or extend the activities - just observe their responses and record their ideas. Systematic investigation will be encouraged during the intervention phase.

Class discussion, as has been mentioned by many of you, allows children to bounce ideas off each other and it's therefore not the most effective way of getting the children's own pre-existing ideas. Instead, we can use other methods which have worked well. Class discussion will no doubt be used during the intervention phase when we are actively encouraging children to reconsider their ideas.

Class Log-Books

These are books into which anyone can make an entry, either written or drawn, and they worked well in the first phase. They are not for investigations/experiments or results, only observations and comments/ideas. If entries can be named (and dated if possible) then the log-book is a valuable record of a child's reaction or observation.

A simple instruction like, "Have a look at these things, and there's a book here for you to write/draw anything you notice", is as explicit as necessary, though enouragement might also be required to maintain interest.

Drawings/Diagrams

These are best done when the children have had a chance to experience the activities for a period.

The main purpose of these is to give children a non-verbal way of expressing their ideas. It's not the drawing skills that are important, it's the ideas. If it's helpful, then children can add explanatory comments to their drawing. Where necessary and practicable, ask individuals to explain their drawings and add the explanation to the drawing yourself, either on the reverse of the paper, or on an attached sheet.

Exploration Experiences

These activities should be set up and left in the classroom for the children to have the opportunity to play with. The activities shouldn't be extended in any way by adding variables or changing conditions, and no measurements or 'results' need to be recorded.

STRING TELEPHONE

Equipment 2 yogurt pots, one knotted onto either end of a 4 metre length of string

An introduction could be something like: ''With a partner, hold one yogurt pot each and make sure the string is pulled tight between you. Take it in turns to whisper a message to each other (e.g. 'draw a house', 'write your name 5 times') and see how well you hear the message. Write down what you think you hear and then see if it was what your partner said.

DRUM

Equipment Drum, or tambourine
rice or silver sand
beater

Instructions, for example: beat the drum and listen to the sound it makes. Put rice/sand on it and watch what happens when you beat it. Put your finger lightly on the drum when it is making a sound.

Draw pictures to show:

(a) how you think the drum makes a sound;

(b) how you think the sound gets from the drum to you so you can hear it;

(c) how you think you hear the sound.

PLASTIC LEMONADE BOTTLE-FUNNELS

Equipment Top of plastic lemonade bottle.

Put the funnel to one ear. Listen to the sounds around you. What difference does the funnel make?

Individual Drawing

Draw how the funnel helps you to hear differently.

RUBBER BANDS

Equipment Rubber band
fingers, pencils etc. to hook band around

Instructions, for example: "Stretch a rubber band between a thumb and a finger/ fingers and a pencil etc. Pluck it with your other hand and see what you notice. Can you play a tune?

LISTENING WALK

If possible, take the children on a walk around the school and/or grounds and get them to listen to the sounds around them.

Class Log Book

A 'Sounds' or 'Listening Book' in which children can write/draw about any of these activities, and also

(a) types of sound which they have heard recently;

(b) sound makers they have experienced recently.

APPENDIX IV

Interview Questions (March and June 1988)

STRING TELEPHONE

1. Tell me what you did.

2. How do you think your message got to your partner? (Probe 'transmission').

3. Supposing you just whispered without the string telephone, what do you think would happen to your message?

4. Why do you think it helps to have a string telephone?

5. What do you think the string does?

DRUM

1. Tell me what you did.

2. What do you think makes the rice move when you bang the drum?

3. Do you think you could bang the drum without moving the rice?

4. How do you think the sound gets from the drum to you so you can hear it?

FUNNEL

1. Tell me what you did.

2. What do you think has made a difference to the sound of the classroom? How?

3. How do you think the sound gets from the classroom to you so you can hear it?

4. What do you think would happen with a bigger/smaller funnel?

RUBBER BAND

1. Tell me what you did.

2. What do you think makes the rubber band make a sound?

3. Do you think you could tell whether the rubber band was making a sound if you couldn't hear it? If so, how?

4. How do you think the sound gets to you so you can hear it?

EVERYDAY SOUNDS

1. What sounds can you hear at the moment?

2. How do you think you can hear these sounds?

3. How do you think you can tell what it is you're hearing (i.e. how can you identify different sounds?)

4. Suppose you're listening out for something, what makes a difference to how well you can hear? (e.g. distance, volume, pitch, background noise)

APPENDIX V

Teacher Guidelines for INTERVENTION (May 1988)

Timespan: April 25th - May 27th (5 weeks)

[Post-Intervention interviewing: June 6th - 24th]

The most important aspect of the Intervention will be the children carrying out investigations based upon questions which they would like to answer.

The children will obviously need help in turning their questions into investigations, and to help them, you and us in this process some starting points for investigations will be given. As before, children should be encouraged to decide on the particular investigation they want to do themselves, though it will be concerned with the activity you suggest to them, in the first instance. Please use these starting points, because it's important that all of the children have experiences of these fundamental concepts before they do more elaborate work. Obviously, some children are likely to be capable of more complex thought, while others will be stretched sufficiently by considering these basic activities. Remember, if children are at the stage of simply making observations then investigations involving imperceptible things are inappropriate.

Helping children to make generalisations

Another very important point is that the investigations are not seen as isolated activities but that children are encouraged to relate them to other experiences they have had. One possible route into this area is to pick up on the experiences which children use to interpret activities - for example, "It's like when you ride your bike down a hill".

- What makes it like riding a bike down a hill?
- Is there anything else that it's like?

Work on developing vocabulary

Children's everyday vocabulary seems to develop to include some words which scientists use to describe events. If children are using words like, for example, 'vibrate', it's important to establish what children mean by them. Try collecting examples of situations when words are used, and also alternative words that different children are using to mean the same thing. This again can't be done in isolation from activities such as investigation - words are used in a context, and **when** words are used is an integral part of the word.

There are certain observations which it would be beneficial for children to have made accurately in order to give them a firm foundation from which to investigate. If possible try to encourage children to pay careful attention to their activities.

To summarise, base your intervention work around children trying out their ideas in investigations.

(1) **Trying out the children's ideas**

Start from simple investigations and pursue them scientifically. Encourage

- predictions
- careful planning
- careful observation
- considered conclusions

(2) **Vocabulary**

This is an integral part of other activities, and should be considered in this context.

- Establish common, agreed definitions for important words which children are using, particularly if they are scientific words.
- Collect alternative words which different children are using to mean the same thing.
- Use the above two to try to encourage children not to use an unsuitable word when you can suggest an alternative, everyday word which is better.

(3) **Generalisations**

This is an integral part of other activities. Look for ways to draw links between the activities and children's experiences, possibly through the way children explain what happens in their investigations/observations.

(4) **Observation**

Use any appropriate opportunities which arise to encourage children to extend their skills of observation, to help them build up their bank of background knowledge and experience.

(5) **Recording**

Please find an appropriate form in which children can record their work. It would be very valuable for us to be able to have this work after the intervention, or at least a copy of it.